Pelican Books

Oil and World Power

Peter R. Odell was born in 1930 in Coalville in the centre of the small Leicestershire coalfield, giving him an early interest in energy problems. He graduated with first class honours in geography from Birmingham University, where he also received his Ph.D. After military service with the R.A.F. and a year studying at the Fletcher School of Law and Diplomacy in Boston, U.S.A. he took up a post in the Economics Division of the Shell International Petroleum Company in London. In 1961 he was appointed lecturer in the Department of Geography at the London School of Economics and seven years later, in September 1968, he was nominated to the Chair of Economic Geography at the Netherlands School of Economics, now part of Erasmus University, in Rotterdam. He is now the Director of the University's Centre for International Energy Studies and since 1982 has held a special Chair in Energy Studies in the Faculty of Economics. He is also a Visiting Professor at the London School of Economics and at the College of Europe in Bruges, Belgium. He has lectured at universities and institutes and has been invited to present papers to seminars and conferences on the geo-economics and the geo-politics of oil and energy in more than thirty countries in five continents. He was adviser on offshore oil policy to the U.K. Department of Energy in 1977/8 and has also advised other governments, international organizations and large international companies on oil and energy questions. He is the author of *An Economic Geography* (1963), *Oil: the New Commanding Height* (1966), *Natural Gas in Western Europe: A Case Study in the Economic Geography of Energy Resources* (1969), *The North Sea Oil Province* (1975, with K. E. Rosing), *The West European Energy Economy: The Case for Self-Sufficiency* (1976), *The Optimal Development of North Sea Oilfields* (1977, with K. E. Rosing), *Energy Needs and Resources* (1977), *The Pressures of Oil: A Strategy for World Economic Revival* (1978, with L. Vallenilla), *British Oil Policy: a Radical Alternative* (1980), *The Future of Oil* (1980, paperback edition 1983, with K. E. Rosing), and *Energie: Geen Probleem?* (1981, with J. A. van Reijn). He has also written a very large number of papers, articles and contributions to books on his special interests in oil in particular, and on energy in general. He is Chairman of the Benelux Association of Energy Economists, European Editor of the *Energy Journal* and a member of the Editorial Boards of *Energy Policy* and the *International Journal of Energy Research*. He is a member of the Royal Institute for International Affairs (London), a Fellow of the U.K. Institute of Petroleum, a member of the International Association of Energy Economists and a Fellow of the Royal Society of Arts.

Peter R. Odell

Oil and World Power

Eighth edition

Penguin Books

Penguin Books Ltd, Harmondsworth, Middlesex, England
Viking Penguin Inc., 40 West 23rd Street, New York, New York 10010, U.S.A.
Penguin Books Australia Ltd, Ringwood, Victoria, Australia
Penguin Books Canada Limited, 2801 John Street, Markham, Ontario, Canada L3R 1B4
Penguin Books (N.Z.) Ltd, 182–190 Wairau Road, Auckland 10, New Zealand

First published 1970
Second edition 1972
Third edition 1974
Fourth edition 1975
Fifth edition 1979
Reprinted 1980
Sixth edition 1981
Seventh edition 1983
Eighth edition 1986

Made and printed in Great Britain by
Richard Clay (The Chaucer Press) Ltd, Bungay, Suffolk

Contents

List of Maps, Figures and Tables

Tables

Preface

The first edition of this book was published in 1970. It emerged at that time from some thirteen years of the author's increasing familiarization with the oil industry, first as an employee of Shell, the world's second largest and the most international of oil companies, and then as a student and observer of its continued growth both in size and complexity. During those years I benefited from innumerable opportunities to learn, through debate, discussion and argument, something of the industry's problems and I would still wish to acknowledge my initial debt in that respect to many people in the industry, in consultancies concerned with oil, in governments and in academic circles not only in London, where most of the first edition of this book was written, but also in other parts of Europe, the United States and many countries of Latin America. Whilst I did not have opportunities to see the oil industry in action in the rest of the world, there were few countries with important oil interests whose problems I was not better able to understand as a result of listening to their representatives talking about them on their visits to London and at conferences elsewhere. In a more strict academic context my greatest debt for help given lay with all who participated in the series of seminars on the economics and politics of the international oil industry that Professor E. Penrose and I jointly ran at the London School of Economics and the School of Oriental and African Studies from 1963 to 1967. The broad general idea for this book, as well as many specific ideas in the original edition, emerged from sessions of that very successful series of seminars at which representatives of governments and oil companies gave willingly of their time.

Since the first edition was completed in January 1970 there have, of course, been the quite normal run of daily changes in the world of oil – in terms of new discoveries and changes in patterns of production and demand. These in themselves have been important enough to make revisions of the text necessary from time to time. In addition, however, there has since 1971 also been a set of abnormal and traumatic developments which have completely changed the traditional pattern of organization of the industry. Thus, the third edition, published in 1974, included a complete new chapter, 'The World of Oil Power in 1974', which made a preliminary attempt to put the changes into perspective and to show how these changes opened up many new options for the future of the industry. Since then the revolutionary change in oil power began to express itself more clearly and by 1975, when the opportunity for publishing a fourth edition arose, it was possible to indicate the lines of a possible longer-term order in the new system. Five years later, things were a little clearer – though the oil world was still beset with much uncertainty in respect of its medium to longer-term evolution. Thus, the sixth edition gave the opportunity for a somewhat more definitive statement on the new order. That was the substance of a completely rewritten Chapter 9 but, in addition, all the other chapters were thoroughly revised and, where necessary, restructured to take cognizance of the way in which the changed situation had influenced attitudes and policies towards the oil industry in the various parts of the world. Two years later, the seventh edition provided another opportunity for a necessary up-dating of developments in world oil in general, and for a definitive version of the revolution in the world of oil power in the first half of the 1970s. This was covered in the retitled Chapter 9. Part of the previous Chapter 9 (much revised), plus additional material on the rapidly changing outlook for oil demand, in the light of a decline in oil use in three successive years (1980–82), and an evolving new organizational shape for the world oil industry, in which process 'international oil' is being replaced by 'a world of oil regions', constituted a new Chapter 10, 'The Response to O.P.E.C.'s Challenge'. In the eighth edition this chapter has been rewritten to incorporate an analysis of changes over the past three years in O.P.E.C.'s role and influence. The rest of the book was initially revised and updated to late 1985 but, given the traumatic changes in the world oil situation and prospects after mid-January 1986, a postscript was added in May 1986, as Chapter 11, to provide a preliminary interpretation of these recent events.

Over the years many people have helped in the preparation of the manuscripts and appropriate acknowledgements have been made in the various editions. For this edition I would like to thank Eibellien Visscher in Rotterdam for retyping parts of the manuscript and Emil van Dijk for updating the illustrations. Continuity in help and encouragement from the first through to this eighth edition has been extended by my wife and children who, as usual, have also accepted many hours of my preoccupation with the problems of oil, rather than with those of the family. My thanks to them and I hope they remain convinced that the result continues to justify their forbearance.

Peter R. Odell, Rotterdam, November 1985, May 1986

1. Introduction: The World's Oil Industry

A description of the world's oil industry demands the use of many superlatives. By any standards it is the world's leading industry in size; it is probably the only international industry that concerns every country in the world; and, as a result of the geographical separation of regions of major production from regions of high consumption, it is first in importance in its contribution to the world's tonnage of international trade and shipping. Because of these and other attributes, such as its involvement in both national and international affairs, a day rarely passes without oil being in the news. Often the significance of such news items is not apparent in isolation or without some background knowledge of the way in which the industry has been and is organized internationally, and of its impact upon individual countries and groups of countries in which its operations and interests lie. This book aims at providing such a background by describing and analysing the oil industry's affairs and relationships around the world.

It is concerned almost exclusively with the period of over forty years since the end of the Second World War, when, except for the last five years, growth in the oil industry was faster than that in most other large-scale economic activities. Already by 1950 twice as much crude oil was produced as had been produced in 1945. Ten years later, in 1960, production had doubled again to 1,000 million tons. The high rate of expansion continued, and it took only five years more for the next additional 500 million tons annual output to be achieved. In 1968, only three years later, another leap of 500 million tons in annual output of a total of 2,000 million tons was achieved and at that time the prospect of an annual output of almost 3,000 million tons by 1975, and of 4,000 million tons by the early 1980s was considered to be as near certain as any forecast could be. Since 1973, however, for reasons which we shall discuss later, the outlook for the future production and use of oil has changed. Annual increases in the size of the industry ceased in 1979 since when production and use has declined and is now back to pre-1973 levels. Even so, there are still possibilities for growth in the size of the industry, though the doubling rate will now be decades rather than less than ten years, as over the two and a half decades from 1948. By 2035 a world oil industry of up to twice its present size is likely to develop. If this is so, then more oil tankers on the roads and railways, new pipelines to major consuming centres in most countries of the world, additional refineries on major estuaries and elsewhere, more tankers for moving oil around the world, and a continued flow of news items about oil exploration and development efforts in hitherto unexpected places such as offshore China, the Arctic, and Antarctic and even under the deep oceans can be expected for as far ahead as we can see.

But such generalizations about the future of the oil industry at the world level are somewhat misleading, particularly as the industry has never been a fully interlocking system and it now seems likely, in view of recent events, to split up even more – perhaps to become, as we shall show later in the book (in Chapter 10), an industry which is essentially regionally organized (see Map 12), rather than international in structure. Meanwhile, there is at the present time, as one would expect from the international division between the communist group of countries and what the Americans call the 'free world' (a description often used in oil company literature), a relatively clear break between the oil industry of the communist nations and that of the rest of the world – though, as we shall see later, there are certain interrelationships between them which do appear to be of increasing significance. Moreover, even within the communist world the deep cleavage between the Soviet Union and its European allies on the one side, and China and some of its friendly neighbours on the other, has also been reflected in a lack of contact between the oil industries of these countries. Until the mid 1960s increases in Soviet oil output were planned on the assumption of an increasing oil import requirement by China. Since then oil trade between the Soviet Union and China has been eliminated and China's recent successes in achieving the expansion of its own oil industry have already made it self-sufficient in oil and now seem likely to make it an important exporter to other parts of the world – particularly Japan – by the end of the 1980s.

In the rest of the world, early in the post-war period, a split emerged between the oil industry of North America (mainly the U.S.A., but including Canada and, to a lesser degree, Mexico) and that of the remaining non-communist world. Before the war and up to 1945, in as far as wartime supply contingencies made this possible, the U.S.A. was the leading exporter of petroleum products to Europe and to other parts of the world. The U.S.A. continued to export oil to Western Europe after 1945 but it soon lost its dominant position as its relatively slowly growing oil production was required at home to sustain the country's rapid economic development. Thus its oil industry became a largely separate entity from the international industry, with such differences in price levels and in organization from the industry in the rest of the world as to necessitate an increasingly autarkic policy on the part of the U.S.A. Indeed, the domestic U.S. oil-producing industry would have been greatly diminished in size had it been subjected to competition from lower cost crude oil from abroad in the 1950s and the early 1960s. This situation, however, changed in the second half of the 1960s because of the seeming inability of the U.S. oil industry, protected though it was, to produce enough oil to meet the country's growing needs. Imported oil gradually increased in importance from its earlier controlled level of about 12 per cent of total supply to more than 30 per cent by the late 1970s. Then, in light of the changed international oil supply situation and higher prices, coupled with weakness in the demand for oil, imports of oil began to decline and domestic production to increase so that the latter was able to meet an increasing proportion of the country's total needs. A reappraisal of this new U.S. situation is made in Chapter 2.

Outside North America and the communist world we have the territory of the so-called 'international industry'. 'International' as used in this context does not, however, imply an industry owned and/or controlled by the world's many nations. Instead, it refers to the fact that the oil industry is one which operates internationally, within the framework of a complex network of relationships which connect most countries of the non-communist world. It was originated and initially controlled by a very small group of companies known, in oil industry terminology, as the 'international majors'. These seven companies (which are described below) were, until recently, responsible for something like 80 per cent of all oil production in the world outside the communist countries and North America.[1] In the same areas they still own or control over 70 per cent of the total refining capacity and they operate either directly or indirectly, through long-term charter, well over 50 per cent of the tonnage of internationally operating tankers. Just a few years ago these percentage figures were even higher.

The companies concerned are international in a number of particular ways: first, in that their operations are world-wide; second, in that they

1. Since 1974 most of the major oil-exporting countries have nationalized these companies' producing facilities. The companies, however, retain much influence over the oil as most of it still has to be marketed through their facilities and sales outlets. One of the 'seven' has, moreover, now disappeared as it has been taken over by another of the companies (see below page 14).

employ nationals of many countries; and third, in that they have locally (that is nationally) registered subsidiaries in many countries of the world. But their ownership and their ways and methods of working are limited to those of just three countries, most especially to those of the U.S.A. Thus, of the 'international majors' no fewer than four have their headquarters and the overwhelming majority of their shareholders in the U.S.A., which also provides all the top management and a high proportion of the lower echelons as well. The largest is Standard Oil of New Jersey, which throughout most of the world traded, until recently, under the Esso ('tiger in your tank') sign – except, ironically, in the U.S.A., where the nationally trading subsidiary was, rather amusingly in the light of the group's strength and size, known as Humble Oil! In 1972, however, there was a corporate decision, taken after considering many alternatives and after much debate by top management, to establish a new worldwide name for the company – the Exxon Corporation. This name has already been generally adopted in the United States and is gradually spreading to other parts of the world.

Another Standard company, Standard Oil of New York (almost every state in the U.S.A. secured a 'Standard' company when the coast-to-coast 'empire' of Rockefeller's Standard Oil was broken up by anti-monopoly legislation early in the century), is the parent company of many subsidiaries around the world trading as Mobiloil or, simply, Mobil. The other Standard company which achieved a place amongst the international 'majors' is Standard Oil of California (Socal), which, after many years of operating abroad jointly with one or other of the other majors, is now 'going it alone' under the Chevron sign. In 1983 it took over Gulf Oil in the U.S. and most of its activities elsewhere in the world. Gulf Oil was one of the original 'seven sisters' (it had its head-quarters in Pittsburgh) but it no longer exists as a corporate entity. It had earlier become a major force in the Middle East when it secured a 50 per cent interest in the extremely prolific Kuwait oilfields in the late 1930s. When it was finally able to develop production from these fields on a large scale after the Second World War, it rapidly diversified into refining and marketing operations throughout Europe and in parts of the Far East and South East Asia. It was, however, terminally weakened by the nationalization of its oil reserves and production in the Middle East. It was in these circumstances unable to operate its remaining activities profitably enough and thus succumbed to Chevron's take-over bid. Finally, there is the Texaco oil company, which, as its name suggests, operates out of Texas. It achieved the reputation of being an aggressive company with major overseas interests in the Caribbean and

South America and with important shares in joint producing companies in the Middle East. It too is involved through a wide range of subsidiaries in refining and/or marketing oil products in some forty countries.

Though these companies have – and, indeed, deliberately foster – different corporate images, they do have one overriding attribute in common: their Americanism. Although most of their subsidiaries appear to have a management consisting mainly of nationals of the country concerned, it is seldom necessary to dig very deeply into their so-called 'organigrams' to find the key U.S. personnel whose job it is to ensure that American professionalism and expertise in oil reaches into the furthest parts of the company's empire. At the same time the U.S. managers have the responsibility for ensuring that the policy of the subsidiary is in line with the authorized interpretation of centrally taken decisions. The essential communications infrastructure of these U.S.A.-centred 'international' firms certainly rivals that of the foreign services of the majority of the world's nations and they have, indeed, been in the forefront of the efforts to secure private global communications by means of orbiting satellites.

This is also true of the two remaining 'international majors' – the Royal Dutch/Shell group and British Petroleum. The former is an Anglo-Dutch enterprise with the larger shareholding interests of the Royal Dutch parent making the Dutch element the more important in the ratio of 60:40 (a division which is also reflected in the 4:3 split of the seven managing directors in favour of the Dutch component). The company's operational and commercial headquarters are, however, in London from where the day-to-day business is run, whilst the head office in The Hague is more concerned with fundamentals such as exploration and production activities, refining and research. The Royal Dutch/Shell group is second only to Standard Oil of New Jersey in size, while the complexity of its operations is probably even greater, given Shell's even wider international commitments – even in areas such as Eastern Europe where the U.S. companies are not represented. There are, indeed, still only a handful of countries in the world in which there is no local Shell company. British Petroleum, or B.P. as it is generally known, is a wholly British-owned enterprise (except for some foreign ownership of its shares) in which, incidentally, the state had a more than 50 per cent holding from 1913 to 1979, except for a short period in the 1960s when it was reduced to 'only' 49 per cent. However, this dominant state holding, much to the disbelief of foreigners, did not result in the British government as such exercising a dominant role in the company. In fact, successive governments over many years elected to exercise no influence at all. They nominated only two directors to the seven-man board and until recently the government-appointed directors did not even 'report back' to the government, let alone seek instructions or advice on how to vote on major B.P. policy decisions. They apparently remained quite free to act as they themselves thought best. In other words the state remained a sleeping partner in the enterprise, content to let B.P. operate just like the other private-enterprise international oil companies. The possibility that this might have been changed after 1974 when the new Labour government came to office with the declared intent of creating a state oil company did not materialize as the government decided, instead, to create an entirely new state oil entity – the British National Oil Company (B.N.O.C.). B.P. was left severely alone. Thus, the state's interest continued to lie only in drawing its considerable dividends from the company, which, like Gulf Oil, had the advantage of access to massive supplies of crude oil from the Middle East in the post-war period and on the basis of which it had both the wherewithal and the incentive to diversify into refining and marketing operations in an increasing number of countries. These eventually included the U.S.A., where, after a series of small incursions to test the market, and in the light of what appeared to be important oil discoveries in Alaska, B.P. made up its mind in 1968 to go into refining and marketing in a very big way. It thus made take-over bids for a number of medium-sized oil companies. Favourable decisions by the U.S. Department of Justice (which is responsible for investigating all take-over bids in the light of the country's anti-trust legislation) gave the green light for the implementation of this major policy development by the company. This has had the effect of making B.P. about the ninth or tenth largest oil company in the United States. It has, moreover, since become one of the country's very biggest producers as a result of its oil discoveries in Alaska. These turned out to be very large indeed and have now achieved full production.

These then are the major international oil companies and, as the figures given earlier indicate, they constituted the dominant elements in the international oil industry until just a few years ago. Before the Second World War they collectively created a cartel-type arrangement, having agreed in the post-depression of the early 1930s on ways to hold their traditional market shares. Effects of the Second World War and post-war U.S. anti-trust legislation then eliminated the possibility of a continuing formal agreement between them, but nevertheless, until the late 1950s – because of their international organization and structure and their effective control over supplies and reserves of crude oil – they were able to make the industry work in the way in which they wanted. The mutual understanding between the companies over supply and pricing policies ensured their ability to earn high profits in this period.

Thereafter, their ability to dominate the industry began to fall away. This change was the result of the growth of new elements in the international oil business. One element was the rapid growth of important state-owned or state-supported European enterprises, such as Compagnie Française des Pétroles (C.F.P.) of France and Ente Nazionale Idrocarburi (E.N.I.), the Italian state company. The interests of these companies did not remain confined to Europe, though this is where they remained strongest. They also spread their activities to many other parts of the world, where they began to participate not only in production ventures, but also in refining and marketing activities.

An even more important new element in international oil arose from the growing international activities of many U.S. oil companies which had formerly had interests only within the U.S.A. but which, since the early 1950s, sought both oil supplies and markets abroad. They did this initially in order to obtain lower-cost crude oil supplies than they could produce from their fields in the U.S.A., with the intention of shipping the foreign oil back to the United States so as to increase the profitability of their domestic refining and marketing operations. The U.S. government, however, imposed oil import quotas in 1959 and thus severely limited the amounts of oil these companies could supply from overseas to the U.S.A. Thus, they had further to diversify their activities geographically by going into refining and selling oil in other parts of the world – notably Western Europe – so that they could secure revenues from the crude oil supplies that they had discovered and developed abroad, often at a high capital cost. In a world of generally booming markets for oil they did not find such expansion too difficult, even though of course, their expansion in such markets was partly at the expense of business that might otherwise have gone to the major companies. Somewhat later, however, as the growth of markets for oil in Western Europe and Japan began to slow down, these companies new to the international business had to face more intensive competition for business. Some of them, in fact, have since been taken over by the 'majors' as a means whereby the latter could reduce the risk of losing any further percentage share of the total markets available.

Neither did most of the companies new to the international oil business show much willingness to get involved with refining and marketing oil in Asia, Africa and Latin America, where the methods of doing business were very different from those in the U.S.A. or Western Europe. They thus chose not to take particular advantage of the increasingly strong reaction in the developing world in the 1950s and the 1960s against the control effectively exercised over most of the oil business in Latin America, Asia and Africa by the seven major international companies, which, through the subsidiary companies they established in the earlier days of the oil industry in most Third World countries, sought to sell crude oil and products from their larger-scale operations in the major oil-producing areas.

The reaction against the international oil companies in the Third World emerged from an enhanced degree of economic and political nationalism in a period in which colonialism in any shape or form has been highly suspect. Many countries of the developing world viewed the control over their oil industries exercised by one or more of the major international companies very much as a form of economic colonialism, and thus incompatible with the status of a sovereign nation. The result was a growing tendency for such nations to introduce increasingly effective controls over the companies concerned through import quotas, price controls, and insistence on the employment of nationals rather than expatriates, together with the imposition of unfavourable taxes and other regulations. Some of the countries concerned even established and encouraged alternative systems for securing their supplies of oil from overseas, and it was in such cases that entities and companies new to the international oil business could find opportunities for cooperation with the countries concerned. E.N.I. and the French state-sponsored entities did so to some extent in Africa and Asia, but the American 'independents' (that is, U.S. oil companies other than the five American 'majors' at that time) only followed suit to a limited degree, before deciding that they also were not particularly welcome in most developing nations.

This nationalist reaction against the major oil companies arose first amongst the larger Latin American nations, which, after a century or so of political independence dating from the mid nineteenth century, also sought their economic independence from what they considered to be an important element of American imperialism. Thus, countries such as Chile, Brazil and Argentina gradually restrained the freedom of the international companies to operate within their territories. The companies were, for example, refused permission to expand their activities beyond those existing at a certain date, or were obliged to integrate their own operations into a framework establish by state control and direction. Moreover, to supplement and intensify these measures the countries concerned often created their own state-owned oil entities which were charged with the overriding responsibility for ensuring the provision of the countries' needs for petroleum. Today, of the fourteen nations of

South America only Paraguay is without its state oil company. This trend towards oil nationalism has also been taken up by many of the more recently independent countries in Asia and Africa, and today over forty countries in these two continents have state-owned oil entities with responsibilities extending from exploration to final marketing. The trend towards state control and/or ownership appears likely to continue amongst the world's developing nations, where suspicions about the wisdom of permitting the international companies to look after the oil sector are still widely held. Such suspicions have, indeed, been enhanced by the fundamentally changed relationships since 1971 between the oil companies and the major oil-producing countries. Direct relationships between some poor oil-consuming countries and the oil-producing countries have been established over the last decade so that the business of the international companies is being squeezed even more effectively. One good example of this is the joint Venezuelan/Mexican agreement (signed in 1982 under the name of the San José Accord) to provide oil to poor countries in the Caribbean and Central America under specially favourable conditions.

Except for their oil exploration and production activities, it was, however, in the countries of Western Europe, together with Japan, that the principal post-war expansion of the international oil companies occurred. Even in these areas the question of allowing these companies freely to supply, refine and market the oil required was raised more and more frequently, even before the oil crisis of 1973–4 and much more so since then, as the countries had to face up to much higher oil import bills. Thus, countries as diverse as Finland, France, Italy, Austria, Spain, Norway and Britain[2] have all decided to place oil partly, at least, in the public sector, through the establishment of a state-owned oil enterprise. And even where state companies have not been established, other European governments have, almost without exception, involved themselves in attempts to persuade the companies concerned to pursue policies in respect of the development and location of refineries, and the price of products, etc., which meet the conditions the country considers nationally acceptable. It is perhaps Japan, as will be shown in Chapter 6, which has taken this sort of persuasion, through supervision and control, to a higher degree than any other country. There, through the powerful Ministry of Trade and Industry, national influence and control is exercised at all significant points over the activities and decisions of the international companies selling oil to or in Japan.

2. Though note that Britain's Conservative government, under Mrs Thatcher, subsequently decided (in 1983) to 'privatize' the British National Oil Corporation (B.N.O.C.) created by an earlier Labour administration.

Thus, since the early 1960s there have been increasingly severe limitations on the freedom of the international oil companies to take decisions solely in the light of their own corporate interests and to exploit opportunities for expanding their activities in refining and marketing oil. They have therefore expanded less rapidly than would otherwise have been the case and have had their degree of dominance in these downstream activities somewhat curtailed. Even in the changed oil world of the 1970s, however, they still remained powerful enough to restrain oil supplies to particular countries from time to time and to secure major increases in oil prices, generally sufficient, indeed, to maintain their levels of profitability.

At the same time these companies have faced even more serious limitations on their ability to do as they would commercially and corporately have liked to have done in the main areas of oil production. From the mid 1950s onwards – and more strongly after the formation of O.P.E.C. in 1960 – all major producing countries began to insist on securing a larger share of the profits from oil production. In this they were increasingly successful over the years. Then most of the oil-exporting countries also began to claim that the companies' operations should either be made part of a joint venture with a state-owned entity, or brought much more directly under the general overall control of a Ministry of Petroleum or a similar official body. Moreover, other non-major companies were often invited into a country to participate in, or even to initiate, an oil development programme, to the detriment of the major companies' share of total production.

Producing countries took these lines of action against the companies which were working oil concessions in their territories in the light of their perception of their evolving economic and political interests. In the early post-war period of rapidly increasing demand for oil, at price levels much in excess of the long-term supply price for oil, the governments were usually easily able to secure their objectives because, no matter what they demanded and obtained, the business still remained very profitable for the companies. In the circumstances of their control over the world's major oil markets the companies could increase prices more than sufficiently to offset the additional payments involved, and so they were able to increase their profit levels at the same time.

However, this situation, which had been highly favourable to the interests of the producing countries, deteriorated after 1957 under the impact of a world surplus of oil-producing capacity and a consequent price weakness in many markets. The producing countries soon found that the international companies, with the flexibility which was given to

20

them by the fact that they had production operations in a number of major exporting countries, were able to concentrate their activities in those countries where there was least interference and the lowest tax demands from the governments concerned. The advantage this flexibility gave to the companies in their relations with the producing countries was a factor in the companies' enthusiastic development of new oil resources in Libya and Nigeria in the 1950s and early 1960s, and in Australia, Alaska and the North Sea in more recent years. The companies fully appreciated that developments in such new areas would enable them to withstand any unwelcome and excessive pressures from the governments of the longer-established oil-producing areas, especially in conditions of a world surplus of oil.

The oil-producing and exporting countries were made painfully aware of this fact of international oil life in the late 1950s when a potential oil-production surplus of serious dimensions, caused by the phenomenally successful search for new oil reserves earlier in the decade, persuaded the companies to announce significant overall reductions in the prices they 'posted' for crude oil as a means of stimulating demand. These price reductions decreased the revenues per barrel of oil produced which host governments received from the oil companies because, in most cases at that time, government revenues were calculated on the basis of the posted prices. The result of this shock to the oil producers was the formation in 1960 of the Organization of Petroleum Exporting Countries (O.P.E.C.), whereby the five countries concerned (viz. Venezuela, Saudi Arabia, Iraq, Iran and Kuwait)[3] sought collectively to enhance their bargaining power by standing together in order to prevent the companies from playing one country off against another whenever there was likely to be a surplus availability of oil relative to expected demand.

O.P.E.C. grew in stature – and size – quite quickly in the early days of its existence and achieved its first success when it prevented further reductions in posted price levels in spite of a very weak international oil market in the early 1960s. This ensured the maintenance of government revenues per barrel of oil produced. It then also successfully negotiated technical changes in the methods whereby company profits in the producing countries were calculated with the result that government revenues per barrel of oil exported were regularly increased by a few cents. Up to 1970, however, O.P.E.C. had not succeeded either in developing a strategy for taking over the power of the oil companies, or in introducing a mechanism for controlling and regulating the output of oil

3. These five founder members of O.P.E.C. were later joined by a number of other oil-exporting countries. See Chapter 4 for more details.

in member countries. In the absence of such controls the potential danger of further reductions in crude oil prices continued to threaten, for new reserves of oil continued to be discovered more quickly than oil was being used, and the industry remained highly competitive, in spite of the high growth rate in demand during the 1960s.

Paradoxically, it was the earlier encouragement that the producing countries had given to companies other than the 'majors' to participate in the exploration and development of their oil that became the major hurdle to O.P.E.C. establishing control over the supply and price of oil. If the oil-exporting countries had only had to negotiate with the seven major international companies, then some form of regulation of oil output, acceptable to both sides, would probably have been achieved in the 1960s. However, the 'majors' were forced to battle for markets with the 'independents' and competition for sales eventually became so acute, even amongst the majors themselves, that the prospect of the international prorationing of oil production – whereby production levels could be tied in to estimated demand levels at a given price (which was the sort of system that worked at the time within the major producing areas of the U.S.A.) – became increasingly difficult to achieve over the first decade of O.P.E.C.'s existence.

Thus, overall cooperation between the producing countries on fundamental steps to change the structure of the international oil industry was slow to develop, and it was not until 1970 that they managed to achieve a consensus for collective action. Finally, at its meeting in Caracas in December of that year, O.P.E.C. declared its firm intention to control the supply and price of oil in the international market. The declaration at that time did not make the headlines as the Organization had not been taken seriously in the West so that its activities and decisions were not then closely followed by governments or the media. Nevertheless, the declaration proved to be the turning point in O.P.E.C.'s evolution because its aims – in respect of the price of oil if in no other way – then coincided with the interests of the world's largest fourteen or fifteen oil companies which, since 1968, had been meeting together as the London Oil Policy Group with the intention of finding ways of eliminating or curbing the weakness in the international oil market. This communality of interests immediately enabled serious negotiations to begin for reversing the downward trend in prices. Governments and companies soon agreed the first increases in posted prices for crude oil – so boosting government revenues per barrel – and these initial steps were followed by a series of rises over the next two years. Finally, in October 1973 the producing countries were able to take advantage of a tight supply situation which had been created by the renewed outbreak of hostilities between Israel and the Arab states, and for the first time were in a position to exercise control over how much oil should be produced. This led to the oil crisis of 1973–4 and to the subsequent rapid evolution of fundamental changes in the international oil system. These changes in the structure of the international oil industry will be described and analysed later in the book.

Thus, today, the international oil industry – the part of the world's oil industry with which this book will be mainly concerned (the exceptions are the chapters on the U.S.A. and the U.S.S.R.) – consists of a number of interacting elements. There are not only the major international companies, such as Shell and Exxon, which formerly ran the industry much as they pleased and which are still very important elements in it, but also a range of additional companies interested in producing oil and selling it on the markets of the world, together with the oil interests of over a hundred national governments, of both producing and consuming countries, throughout Europe, the Middle East and the developing world, and of the international organizations which represent them.

In spite of the recent important changes in their relationships with the oil-exporting countries, the large companies continue to dominate both the stage and the play as the main producers, refiners and sellers of the world's oil. In these roles they are brought into continuing daily contact with the governments of those countries in which they both choose and are allowed to have interests. To some degree these companies also have to act as intermediaries between the conflicting interests of producing countries on the one hand, and consuming countries on the other. Thus, they operate not only as commercial enterprises attempting to maximize their profits, but also as diplomatic channels attempting to keep the oil flowing around the world. Their function in this latter respect may, however, now be a relatively short-lived one, for, if state interests and direct government involvement in oil continue to expand at the rate at which they have expanded in the post-war period in general and over the period since 1973 in particular, then, in the not too distant future it seems more likely than not that the oil companies will no longer be required as intermediaries. The economic and political negotiations needed to get oil from the points at which it is found and produced to the points at which it is to be refined and consumed will then have to be undertaken by the governments of the countries directly concerned. This possibility and the problems to which it leads are examined in more detail in later chapters.

If and when this happens, the 'international oil industry' could become effectively, rather than nominally, international, with all that this implies, both favourably and unfavourably. On the one hand, there are the dangers that could arise as a result of possible confrontation between conflicting national interests regarding oil. On the other hand, however, there are also the opportunities that such a situation would present for the effective internationalization of this most important internationally-traded commodity. This could conceivably be under the aegis of an agency which has overall responsibility for the production, refining, transporting and marketing of the oil within the framework of the most rational system that it is humanly possible to develop. The basis of such a system already exists in the increasingly sophisticated information and logistical systems that have been developed by the separate international companies for optimizing their own production, transport, refining and distribution systems across the frontiers of the many countries in which they operate. This has been done not only for the sake of the companies' own profitability, but also because their managers have continually responded to the intellectual challenge of developing such approaches to the effective organization of the world's largest industry. There seems no reason why this challenge could not, in the foreseeable future, apply also to the even greater problems of establishing an overall, rather than a company by company, optimal pattern of international oil-industry organization. Events in the world of oil power since 1973 have certainly created the opportunities, and the need, for such a development. Those events have, however, also set to work forces which could eventually lead to the dismemberment of the 'international' oil system and its replacement by a number of largely 'independent' regional oil systems with much in common with those of the United States and the Soviet Union at the present time. The latters' oil systems – and their relationships with the oil industry in the rest of the world – are considered in the next two chapters; while the potential regionalization of the international oil system is discussed in Chapter 10.

2. The U.S.A. and World Oil

It is generally accepted that the oil industry was born in the United States (in Pennsylvania in 1859) and that, except for a few years in the latter part of the nineteenth century, the United States had, until recently, always been the world's largest oil-producing, refining and consuming nation. It remains the world's largest oil-consuming nation by far, but it was overtaken by the U.S.S.R. in terms of production and refining over a decade ago. Even so, American oilmen – and many of their fellow countrymen – look upon the oil industry as an American 'invention' and continue to express scepticism as to the ability of any other nation to deal successfully with oil, and over the possibilities of organizing and running the industry in any way other than that tested and tried in the U.S.A.! These attitudes emerge, of course, from the early rise to importance of the U.S. domestic oil industry and the growing overseas interests of the largest American firms, which, as we saw in Chapter 1, have dominated the international oil business since its earliest days. Until recently, only the Anglo-Dutch Shell group and the British Anglo-Iranian (now B.P.) companies provided effective competition at the international level, while Shell was the only foreign company prior to 1970 to make any significant contribution to the development of the oil industry in the United States itself.

Shell has had very important American interests, in the form of the U.S. Shell Oil Company and its subsidiaries (originally 50 per cent owned by Shell International but then taken over completely in 1984 when the remaining 31 per cent of shares in private hands were purchased by Shell International), for a very long time. This American component in the international group has now become so large that it contributes approximately one-third to the total world-wide Shell group's revenues and well over one-third of the group's profits. Moreover, its American expertise in all branches of the oil industry is utilized in the international activities of Shell in a variety of operations. This process culminated in 1967 with the appointment of the president of Shell Oil to a managing directorship of the Shell Group. On the other hand, for non-American Shell Group employees, periods of service in its U.S. operations appear to have been a prerequisite for 'getting to the top'. Almost all those who later became managing directors of the Shell Group in recent years were sent to work in the U.S.A. at earlier stages in their careers with the company. Since the 100 per cent take-over of Shell Oil by Shell International, operations in the United States have been more closely integrated with the Group's international organization and structure (while Shell remained a partly separately owned company U.S. anti-trust legislation made such integration – or even cooperation – illegal). Thus, for example, Houston, the headquarters of Shell Oil, has now been made Shell International's western hemisphere head office – superseding Curaçao, in the Dutch Antilles, which formerly fulfilled this role.

B.P. has also become increasingly involved in and committed to the United States over the past fifteen years. Through exploration and development activities in the United States and through the acquisition of American oil and petrochemical companies, it is now also concerned with all sectors of the oil industry in that country. It was, indeed, B.P. which initiated the exploration and development of the oil and gas reserves of Alaska – the most significant development in the U.S. oil industry in the post Second World War era. Moreover, in the Shell style indicated above, B.P.'s managing directors are now also being appointed from among its senior employees with American experience.

Thus, technically and organizationally, and also from the point of view of the provision of oil-industry machinery, equipment and contractual expertise, the industry remains highly American-orientated – and, as we shall see later in the chapters dealing with the developing countries, this situation often led to charges in such countries that the oil industry was one of the main agents of economic imperialism or colonialism. From the U.S. standpoint, however, the involvement of the American oil industry in the developing world – as, indeed, in Western Europe and other parts of the industrialized world – is presented as evidence of the willingness to make American expertise available overseas, through investments by U.S. oil companies, to all nations that care to make use of it in order to develop and sustain successful oil activities. And the U.S. government has, of course, traditionally considered it a national responsibility to offer protection to this investment and has usually reacted quite predictably to any apparent 'threat' against it by foreign governments as, for example, as recently as October 1980 when the United States took action to prevent the closure of the Straits of Hormuz as a consequence of the Gulf War between Iran and Iraq, and in its imposition of economic sanctions against Libya in 1982 partly as a result of Libya's treatment of U.S. oil companies involved there in exploration and production. Such attitudes and actions in their turn confirm other countries' interpretations of the 'economic imperialism' component in American oil interests overseas. The traditional U.S. view in this respect was well expressed as long ago as 1968 by the U.S. State Department's former Director of the Office of Fuels and Energy. He wrote:

The U.S. government exercises vitually no power of control over the operations abroad of American oil companies. Its concern with them is of another kind. Our companies work abroad in close relationship with host governments. In countries where crude oil is produced the relationship is that of a partnership between company and government, with the company providing the capital and taking the risks inherent to exploration and development in return for a grant by a government of a right to use the resource. In consuming countries oil companies provide the services of refining and distributing petroleum products. Companies obtain profits from these operations commensurate with the large amounts of capital required and the risks assumed. Governments in turn receive an agreed share of profits in return for the rights they have awarded for the resource itself, or revenues by way of taxes accruing from the refining and distribution of petroleum products. These relationships are mutually beneficial to companies and government, and each is dependent upon the other in organizing a system for the production and distribution of petroleum which is economic, efficient and insures a fair return to both, for what they have contributed to the enterprise. The U.S. government is greatly concerned that American oil companies abroad should be able to continue their operations overseas within a framework of mutually agreed relationships with host governments, serving the public efficiently wherever they are, and earning whatever is properly theirs because of the capital, skill and good fortune that have accompanied their operations.

This description of the organization of a large part of the international oil industry indicates how, for a long time (since the 1920s and even earlier in some countries), the oil industry has been part and parcel of the U.S.A.'s world-wide interests. The investments of U.S. private oil companies in virtually every country of the non-communist world have been linked to and supported by official U.S. government policy. Until the post-1974 spate of nationalization of the oil industry in most O.P.E.C. countries, it was estimated that a total of over \$10,000 million was invested abroad by U.S. oil companies. This accounted for about one-third of total U.S. foreign investment at that time and, as the State Department Director of Fuels and Energy observed, 'the loss of a significant part of the investments would be a serious matter to the nation as well as severely inequitable to the private owner'.

The potential loss from a national point of view arose partly from the contribution that oil companies' operations overseas have made to the

U.S. balance of payments. In 1973, the last full year before the process of nationalization of oil production in the O.P.E.C. countries began, it was estimated that U.S. oil industry earnings abroad sent back to the U.S.A. were more than $2,500 million. This exceeded by over $500 million the sum of the cost of the United States' oil imports and the outflow of money from the U.S.A. to finance the continued development of the oil industry overseas. The importance of this to a country with persistent balance-of-payments problems, arising from its massive military and aid commitments overseas and the declining competitiveness of much of its industry, was, of course, always self-evident, but it has now been even more effectively demonstrated by the post 1974 deterioration in the U.S. balance of payments. This has arisen not only out of the higher cost of oil imports, as a consequence of the need for more imported oil at much higher prices, but also from the loss of overseas earnings by U.S. oil companies consequent upon the nationalization of their activities in most O.P.E.C. countries in recent years.

But apart from the narrowly commercial considerations there are wider issues involved in U.S./international oil industry relationships. U.S. governments have been – and remain – concerned with three such wider issues: first, with the security of international oil supplies both for itself and its allies; second, with the economic strength of all nations not unfriendly to it (and this, in part, is a function of oil questions); and third, with 'selling' its own way of doing things for ideological reasons and in this respect too the way in which the oil industry is organized and operates is an important consideration.

The question of the security of oil supplies from overseas will be dealt with later in the chapter when we look at the U.S. position in world oil trade. It is sufficient at this point to indicate the official U.S. view that security of supply must, in the words of the previously quoted state department official, be based 'on the maintenance of a type of relationship between foreign companies and local governments similar in its general framework to that which exists today' – meaning, of course, quite simply a belief that only American companies operating overseas can be trusted to ensure the continuity of supplies to the United States! Given this view, U.S. foreign policy, in areas where crude oil production became important as a result of the activities of U.S. oil companies, had to be concentrated on maintaining the *status quo*. Any threat to this needed to be met with appropriate diplomacy. The way in which this view of the oil world worked can be seen in the case of the United States' relations with Venezuela over the period of about thirty-five years to 1973.

Venezuela's oil was developed by U.S. companies from the 1930s on and it was the major source of foreign oil imported into the United States from 1948 – at the beginning of the period in which the U.S.A. became a net oil importer – through to 1974, when Venezuela's exports became unable to keep up with the growth in U.S. crude oil import requirements. Venezuela's exports to the U.S.A. at first grew rapidly, until the imposition of U.S. import quotas in 1959. Thereafter, they continued to grow slowly as they faced increasingly effective competition, in the now statutorily limited market, from oil imports from the Middle East, Libya and Nigeria. Venezuela's oil development was thus directed much more specifically to meet U.S. requirements than that of any other country; and, up until 1973, more U.S. oil capital had gone to Venezuela than to any other country. Much of this investment took place between 1940 and 1957 when, except for a short period of political difficulty in 1948 (see pp. 78–9), the investment climate for U.S. oil capital was exceedingly favourable. Although changes were made from time to time in Venezuelan taxation regulations and in various other aspects of the concession arrangements, there was no question throughout this period of the nature of the arrangements being fundamentally altered. Indeed, in 1956–7 Venezuela's President, Pérez Jiménez, chose to auction off extensive new areas to foreign (mainly U.S.) companies on conditions roughly similar to those under which the existing concessions were being operated.

In 1958, however, following the overthrow of Pérez Jiménez, the Venezuelans elected to office a left-wing party, Acción Democrática. This party had pledged, while in opposition in exile, the nationalization of the country's oil resources and industry when it achieved power. Its leader, and later the country's new President, following the overthrow of Pérez Jiménez, was Rómulo Betancourt, whose book *Venezuela: Política y Petróleo* (F.C.E., Mexico, 1956) had made out the case for the nationalization of the country's most important and most powerful economic sector. The companies, as in 1948 at the time of the first and very short-lived Acción Democrática government, immediately saw their fundamental interests at stake and it seemed as though they might again react to protect them by seeking to intervene in Venezuela's affairs; if necessary by supporting an attempt to overthrow the government by force. In doing so, they might, under normal conditions, have expected the backing of the U.S. government, acting to protect American oil investment and the nation's oil supplies, in line with its declared policy. The U.S. government undoubtedly made it quite clear in private to both the Venezuelan government and the oil companies that action along these lines would be the ultimate sanction to protect both U.S. investment in Venezuelan oil and the flow of oil supplies to the United States. Such intervention, which would, if necessary, have involved the use of force, would certainly not have been out of keeping with the U.S.A.'s normal practice in the Caribbean area – and in this case with somewhat more justification than in a whole series of other Caribbean interventions by the United States. The oil companies concerned, however – all American with the single exception of Shell – received immediate advice (tantamount, of course, almost to instructions) to work for a *modus vivendi* with the left-wing government. At the same time the latter was persuaded to take no action, such as expropriation of the companies' assets, which would bring the situation to a head. Thus a crisis between the two countries at that time over the oil industry's position in Venezuela was averted, much to the relief of the U.S.A., faced, as it then was, with the difficulties in the Caribbean caused by the defection of Cuba to the communist camp.

It was, indeed, the recently developed Cuban problem which so exercised the U.S.A. at the time. Given that problem, Venezuela was, seen as the key to political stability in the whole region, as well as a test case for the ability of a left-wing, but reformist, government to survive the economic, social and political pressures of opposition from extremists. A crisis in Venezuela had to be avoided and, with the support and guidance of the U.S.A., both the government of Venezuela and the oil companies with concessions in the country pulled back from the brink and eventually evolved a mutually acceptable working relationship which survived for almost another twenty years. The agreement gave the Venezuelan government high, guaranteed, and gradually increasing revenues from oil, so that it was not only able to finance its programme of economic and social reform, but also had some cash to spare to keep potential power groups, such as the army officers, from trying to take power in the country. At the same time the companies retained the ownership of their considerable assets and their ability to continue to make profits on their Venezuelan operations. For the U.S.A. this represented success indeed, not only because it left the reformist government of Venezuela with a chance of succeeding, but also because it secured the single most important source of its oil imports and maintained the essential framework of relationships between the Venezuelan government and its own oil companies. (Chapter 4 discusses this issue from Venezuela's point of view and deals also with events since 1973 which have fundamentally changed the relationships which were established in the late 1950s.)

Such U.S. political intervention overseas in connection with oil has certainly not been restricted to the Western hemisphere, but elsewhere in the world it is very difficult to disentangle the role of oil from other considerations. Collectively, U.S. investment in the oil industry in the Middle East has exceeded that in Venezuela, and the Middle East also increased rapidly in importance as a supplier of crude oil to the United States from the beginning of the 1960s. In the light of these facts it was not surprising that U.S. strategy towards the area aimed at the creation of a collective defence organization against any possible incursion by the Soviet Union, which, of course, has a common frontier with one of the main oil-producing nations of the area, Iran. United States diplomacy in the early 1950s thus worked hard for the formation of the Central Treaty Organization – CENTO – which had as one of its objectives the protection of the American dominated oil-producing areas in the Middle East against external intervention. This diplomatic effort eventually succeeded, when the United States persuaded Turkey and Pakistan, together with Iran, to become the local member nations of the Organization. Although the U.S.A. undoubtedly breathed a little more freely after CENTO's formation in 1955, it should not escape notice that all the main oil-producing nations of the region, with the single exception of Iran, declined to join it. This was basically because at the time the countries concerned felt more exposed to the dangers arising from the existing economic and political interests of the U.S.A. than they did to the dangers from the U.S.S.R. The latter seemed somewhat hypothetical, given that the U.S.S.R., with its own rapidly developing oil production, had less motivation than the United States to become involved in the region.

At that time moreover – and, indeed, until very recently – the U.S.A. remained powerful enough to back up its interests with sufficient military force, or potential force, to ensure that no fundamental change in the favourable situation for U.S. investment in the region occurred. Amongst the devices which the United States used to implement its policy of safeguarding its oil investment and of denying the oil to a possible adversary were first, U.S. backing for the British presence in the Middle East, particularly in the Persian Gulf; second, a supply of arms to favoured rulers; third, intervention with force when necessary – as in the case of the landing of U.S. troops in Syria and Lebanon when there seemed a real threat to the important pipelines passing through these countries on their way from Saudi Arabia and Iraq to the Mediterranean; and fourth, the strengthening of U.S. influence over Saudi Arabia.

U.S. policy towards Middle Eastern oil in the 1950s and the 1960s was thus essentially an intensification of its policy between the wars, when government support and backing of many kinds were used to carve out commercial oil interests for the U.S.A. in an area previously dominated by Britain and France – the successor powers to the pre-1918 Turkish domination of the area. From the 1920s U.S. pressure gradually increased so that it finally became sufficiently strong to force Britain and France to concede ground. As a result U.S. companies secured undivided control over the oil resources of Saudi Arabia and partial control over exploration rights in all other territories on the western side of the Persian Gulf. It was only in Iran that the U.S.A. failed to secure a share of the country's oil rights in the period before the Second World War, but even this was an omission which it was able to make good shortly after the war, in the agreement which followed the dispute between Iran and the Anglo-Iranian Oil Company. In this dispute, which began in 1951, all the oil company's installations were expropriated by the Iranian government and for three years virtually no Iranian oil, other than that required for domestic use, was produced or refined. It was U.S. mediation which eventually, in 1954, produced a formula whereby the dispute was ended. As part of this formula a consortium of foreign oil companies was established to work the Iranian oilfields and the Abadan refinery on behalf of the National Iranian Oil Company. In the light of a more than twenty-year-old American interest in securing dominance over Middle Eastern oil it is perhaps hardly surprising that the agreement involved a 40 per cent interest in the consortium for a group of American oil companies and thus brought the U.S.A. into the one area in the Middle East from which it had previously been excluded.

American successes in expanding and intensifying its oil interests in the Middle East and in keeping out the Soviet Union by means of collective defence agreements and its military strength have, however, been tempered by setbacks arising from change within the area itself – change which the U.S.A. has had to struggle to slow down, let alone stop. In part this has arisen from increasing nationalistic resentment against the strength and dominance of the American oil companies in the region; and in part it emerges from Arab suspicions of the role of the U.S.A. in supporting Israel. The latter consideration has caused some continuing difficulties, such as Arab unwillingness for the Iraq Petroleum Company's pipelines from the Northern Iraq fields to Haifa to be used (see Map 4), and Arab insistence that Israel be denied oil supplies by any company wishing to continue to operate in Arab countries. The more serious repercussions, however, have come intermittently with the Arab–Israeli wars of 1956, 1967 and 1973. On each occasion U.S. involvement on the Israeli side seemed clear to the Arab states and counteraction was thus taken against the United States and/or the American oil companies. In 1956 the partly American-owned pipelines to the East Mediterranean terminals in the Lebanon from the Northern Iraq oilfields were blown up in Syria, and severe restraints were placed on American oil companies' commercial freedom to sell oil to whom they pleased. It took many months before the situation reverted to the *status quo ante bellum*.

In 1967 the threat to U.S. oil interests seemed to be even worse, with a possibility at an early stage of widespread government intervention or even expropriation of their assets. The potential enormity of the repercussions of such action on the oil-producing countries' revenues was, however, quickly appreciated and the Arab nations contented themselves by declaring that Arab oil should not move to the U.S.A. – decisions which, as things turned out, were effective for little more than a few days as far as most major producers were concerned and for no more than a few weeks for the rest. During this period the U.S. oil companies working out of the Arab oil-producing countries continued to supply oil to the non-embargoed destinations, in most of which, of course, U.S. oil interests were involved as refiners and marketers. Meanwhile, the Iranian Oil Consortium, in which, as we have seen, American companies had a 40 per cent interest, took over the responsibility for ensuring that adequate supplies were available for the U.S.A. itself. The disturbance to supplies thus turned out to be a relatively limited one, even as far as the United States itself was concerned. The closure of the Suez Canal (as in the 1956 crisis) did not much affect the flow of supplies to the United States as these were able to go around Southern Africa at little extra cost. Paradoxically, the Arab states' action did, however, permit the U.S. oil companies to earn higher profits as a result of the increase in the rates they could charge for their tankers, and the increased prices they were able to charge for their oil in the tight market conditions which developed because of the temporary supply difficulties.

By 1973, at the time of the next Arab/Israeli war, the United States had significantly increased its reliance on oil imports from the Middle East and it was by then expecting to become increasingly dependent on Middle East oil. The Arab oil-producing nations interpreted this as implying a greater ability on their part to bring pressure to bear on the continuing U.S. pro-Israeli policy and so, following the renewal of hostilities, they announced a complete embargo on all oil sales to the United States. Once again the blow was softened by the ability of the international oil companies to switch their supplies of oil from non-Arab countries to the United States (in exchange for more Arab oil to the non-embargoed countries), but on this occasion, given the greater dependence of the U.S.A. on oil imports than previously, the embargo did lead to supply difficulties in certain parts of the United States and a need for voluntary – and a little compulsory – constraint on oil use in the winter of 1973–4. A much more important result of this experience, however, was an American resolve to try to reverse its earlier acceptance of the idea that it could rely at least to some degree on energy from the outside world. Within a few weeks of the Arab nations' action President Nixon announced 'Project Energy Independence 1980'. This was quickly backed by Congressional support for the finance which was thought to be needed for the effort by the United States to become largely self-sufficient in energy by the end of the decade. General political problems in the U.S.A. (consequent upon the Watergate affair), however, coupled with economic, political, environmental and technical problems associated with the production and the pricing of energy in the United States, made the achievement of self-sufficiency impossible by 1980. Indeed, the United States became even more dependent on imported oil in the period between 1973 and 1979, so raising a set of problems to which we shall return later in the chapter. In the meantime, however, United States' support for Israel has remained virtually as strong as it has always been. Yet, in spite of this, it was U.S. intervention, rather than that of Western Europe (the countries of which, given their greater dependence on the continued flow of Arab oil, sought to appease the Arab nations by moderating their views on Israel), which lay behind the first real move towards *rapprochement* between Egypt and Israel in 1977, and so opened up the possibility of a long-term settlement of the Arab/Israeli dispute – a settlement which the United States saw as one of the two critical issues likely to influence the longer-term future of oil supplies from the region. The second, and more important, issue is the way in which relationships between traditional and revolutionary Islam emerge.

Other changes within the Middle East itself have had a serious effect on those U.S. interests which were established through the oil concession system. Some of the early oil concession agreements, and even some drawn up in the post-Second World War period, virtually gave the U.S. oil companies sovereignty over the territories concerned. In essence, they gave the companies the right to explore without let or hindrance in any part of the national territory concerned. Then, if they discovered oil resources, the oil companies had the right to determine unilaterally whether to exploit them or not, and the extent to which this would be done. And all this was originally conceded to the oil companies by the countries concerned in return for a small royalty payment and some share of the profits, the determination of which was in any case a function of the ways in which the companies organized their accounts. In the changing post-war political and economic climate – and in the light of changes in U.S. oil-company/host-government relations that had already taken place, or were in the process of taking place, in Mexico, Venezuela and Indonesia for example – the U.S. government could not maintain such relationships on its companies' behalf, no matter how satisfactory it might have found the existing situation.

The first major potential crisis occurred as early as 1949 when Saudi Arabia claimed, as part of a revision of its concession arrangement with the Arabian–American Oil Company (Aramco), 50 per cent of the company's profits from selling Saudi Arabian crude oil. Further discussion on the background to this dispute is contained in Chapter 4, but here it should be noted that it was an unwelcome development from the point of view of the U.S.A., which was anxious both to maintain stability in the Middle East – and so needed to assuage the demands of the Saudi Arabian government – and also to ensure the continued profitability of Aramco, its single most important overseas oil entity. The company and the Saudi Arabian government were obviously near to an impasse in their negotiations over the claim when, in order to protect its underlying interests, the U.S. government offered a way out. It suggested a formula whereby Aramco would pay the Saudi Arabian government the 50 per cent of its profits as demanded, but the member companies of the Aramco consortium (namely Standard Oil of New Jersey, Mobil Oil, Texaco and Chevron) would then, in effect, be reimbursed by being allowed to offset these tax payments against their tax obligations in the U.S.A. itself. The Saudi Arabian government thus gained its objective of significantly higher revenues. Net profits for the Aramco companies remained the same. The U.S. government, however, had to forgo the taxes previously paid to it by four of America's largest oil companies from the latters' most profitable operations worldwide.

In other words, the U.S.A. considered its interests in the Middle East oil industry and the way in which it was organized so vital that it was prepared to forgo, on behalf of its taxpayers, the large amount of taxes it had formerly been paid by the companies concerned. This new tax formula was thereafter applied to all other American oil operations overseas. Companies involved in such concession-type oil operations in foreign countries, where local taxes on oil produced and exported were high, thus secured zero tax obligations at home on profits earned from such operations. The oil companies have continued to enjoy this favourable situation ever since, though the whole question has been investigated both by Congressional committees and the Inland Revenue

Service in an attempt to limit the loss of tax revenues. A change was made even more necessary as a result of the greatly increased profits made by the oil industry since 1973, and by the new arrangements between the companies and the oil-producing countries in the context of which taxes, in the normal way in which the word is used, can no longer be said to be paid by the companies. The situation has still not finally been resolved, but it has already been modified to some degree and is likely to be modified still further – to the disadvantage of the companies.

While the U.S. taxpayer can be said to have been consulted 'retrospectively' over 'his' decision to forgo tax in favour of the oil companies and the oil-producing nations (in that the original decision has not been reversed, following elections in which the politicians responsible put themselves at risk), the same cannot be said for the oil consumers and taxpayers of other nations who have also, indirectly, been adversely affected by the decision. This was because the U.S. action helped to strengthen both the oil companies and the oil-producing countries in their relationships with oil-importing countries. Moreover, the adverse effects on these unconsulted entities have added up, of course, to produce significantly adverse effects on the economies of other nations. Thus one can argue that the U.S. government's action, which emerged from collusion with the two other parties concerned – the oil companies and the oil-producing governments – produced a swing in the geo-political relationships of nations. For oil consumers in other nations, the short-term effect arose from the consequential impact which the change in the tax arrangements had on the stabilization of the price structure – or, more particularly, on the posted price structure to which the eventual price of oil products to the majority of consumers was, at that time, firmly related. Thus the introduction of a more flexible and, from the consumers' point of view, a more favourable pricing system was delayed. And as far as taxpayers in other countries were concerned, the arrangement ensured that the companies which supplied them with their oil – and usually refined and distributed it as well – would choose to show their profits in their producing operations in the exporting countries, as this gave them the opportunity of off-setting their liabilities for paying taxes in the United States. This, of course, produced a situation in which the producing companies in the oil-exporting countries so arranged the prices at which they 'sold' oil to their sister companies in other countries that the latter apparently showed very low, or even no, profits on their local refining and/or marketing operations. Thus, all other taxpayers in these countries had to be more heavily taxed in order to make up for the taxes which the oil companies would otherwise have paid. So, for the oil-consuming countries, particularly those of the developing world, consequent high oil import prices, combined with the absence of revenues from the local operations of the international oil companies, produced an economically disadvantageous position. This is one example of the way in which U.S. oil policy decisions have adversely affected other nations, some of which have then had to be given aid in order to make them economically viable, so that the American taxpayer pays twice over. The effect of the U.S. tax concession (a concession which later had to be extended by the U.K. and Dutch governments to the British and Dutch-based international oil companies so that they would not be uncompetitive with their American competitors) in the Middle East and other major oil-producing and exporting countries did, of course, have the desired effect as far as American interests in the international oil industry were concerned. It avoided a showdown between the companies and the governments. A potential major cause of political unrest in the Middle East was eliminated and any long-term interruption of international oil supplies, on which, as has already been indicated, the U.S. was becoming increasingly dependent by the 1950s, was made much more remote.

Thus until the beginning of the 1970s, through a combination of luck, good judgements, appropriate diplomatic pressure, military threats and displays of force when necessary, together with a realization by all parties (U.S. government, U.S. companies and the governments of the oil-producing countries) that everyone's interests were being reasonably well served, at least in the short term, by keeping the oil flowing freely, U.S. companies were able to continue to expand their oil-producing activities in those parts of the world in which they chose to operate. These activities, moreover, came to be generated not only by the large international companies but also, as already indicated in Chapter 1, by a dozen or so other American companies which had, until the middle or late 1950s, been largely confined to domestic U.S. operations.

The motives for the corporate decisions by these companies to expand their operations beyond their familiar territory of the U.S.A. – and sometimes Canada – arose in the first instance from the opportunities which they saw for highly profitable operations in a post-war world short of energy and in which high prices were willingly being paid for low-cost oil from Venezuela and the Middle East. By the early 1950s this motivation was enhanced by the realization that American oil production was being kept in check by institutional factors, the most important of which were the very rigid controls over production exercised by the regulatory bodies such as the Texas Railroad Commission. So growth in the U.S. oil market would necessarily have to be met more and more by supplies from overseas, where the immediate flush of post-war exploration and development had quickly revealed a supply which was not only very low-cost compared with U.S. oil production, but also, in relation to the world's total requirements at that stage, a supply which could only be described as limitless. The domestic companies were thus faced with the alternative either of importing part of their U.S. crude oil requirements through the old-established international groups or of going out to seek concessions overseas for themselves. They decided, in the main, given their evaluation that oil would be made available to them by the major companies only at prices well above the costs of producing and shipping the oil to the U.S.A., that the second procedure would be more advantageous to them, particularly in respect of the higher profits they were likely to be able to earn on foreign operations.

Thus first Venezuela, then the Middle East and, more recently and most significantly, North Africa were searched for possible concession areas by such companies. Many of the companies were quickly very successful in discovering new oil reserves. For example, one of the first companies to take the overseas plunge – Sun Oil of Philadelphia – was soon successful in its development of the concessions which it secured in Venezuela, where it found fields capable of producing over five million tons of oil a year. More recently Occidental Oil of Texas found, without much difficulty, one of the largest fields in Libya and immediately planned to achieve an annual production capacity of 50 million tons (an attempt which was frustrated first, in 1971, by Libyan controls over the level of production and, in 1973, by Libya's expropriation of the company's operations). Not all the efforts of the former U.S. domestic companies achieved such instant or significant successes, but almost all of them found some oil (or gas) and many of them started to ship it back to the U.S. – import quotas permitting – as quickly as possible.

It had become apparent, even before the end of the 1940s, that the combination of oil-producing restrictions in the U.S.A., the rapidly growing supply potential in Venezuela and the Middle East, and the successes of U.S. oil companies in finding oil overseas was undermining the position of the U.S.A. as an oil producer and exporter. U.S. exports had, indeed, dwindled rapidly in the post-war period, reflecting both increased pressure of demand at home on a controlled supply, and the availability of lower-cost oil from elsewhere which could be substituted for American oil in overseas markets. At the same time oil imports into those parts of the U.S.A. most remote from the indigenous producing areas – notably California and the north-east coast – started to grow. By 1948 U.S. oil imports exceeded exports and, for the first time in the history of the oil industry, the U.S.A. became an oil deficit nation. Thereafter, imports continued to grow and, with them, concern for 'security of supply' – the most potent argument in the armoury of weapons used by the domestic oil producers, whose protests, in the face of overseas competition from inherently lower-cost oil, became increasingly vehement.

As this phenomenon of an increasing dependence on oil imports coincided with the political difficulties of the Cold War, and a feeling in the U.S.A. that much of the world was becoming increasingly hostile to its power and influence, politicians were ready to support the security-of-supply argument and pleaded for restrictions on imports of oil. As the pressure mounted the Eisenhower administration was forced to take action. In 1954 and again in 1958 it called for voluntary restraint on the part of importing companies. The larger companies heeded the warning, but there was legally nothing to stop other companies, and particularly those bringing new overseas fields into production, from taking advantage of such voluntary restraints on the part of others – thus carving out larger markets for themselves.

As a result the 'more patriotic' oil companies now joined forces with the other pressure groups against oil imports to get something more than voluntary restraint, which, even with the help of tariffs on both crude oil and oil products, was having no apparent effect on controlling the rate of growth of imports. As a result President Eisenhower introduced mandatory quotas in 1959 on both crude oil and oil products and effectively closed the U.S. market to the unlimited entry of oil from the rest of the world (except for Canada and Mexico, whose oil was excluded from the restrictions on the grounds that imports from these countries were not at risk as they did not depend on ocean transportation). The quotas were related to the U.S.A.'s total use of oil and were set at the percentage level that imports contributed to U.S. oil supplies in 1959 – approximately one-eighth. This meant that the amount of oil imported could continue to grow, but no more quickly than the overall rate of growth of U.S. consumption. This was approximately 3 per cent per annum compared with a 15 per cent annual rate of growth in oil imports over the previous ten years. Thus domestic oil interests were not only guaranteed almost 90 per cent of the existing market but they also secured almost 90 per cent of all incremental demand. External sources of supply, on the other hand, which by 1959 were sending over 50 million tons to the U.S.A., could now look forward to no more than a very modest rate of increase in this market, compared with the high rate of growth that they had enjoyed over the previous decade.

The economic effects of this decision by the U.S. government had far-reaching consequences both inside and outside the country. Inside the U.S.A. it gave a high degree of protection to domestic oil (and other energy) interests. U.S. oil and coal output was maintained at a level far higher than it would have been with continued unrestricted competition from foreign oil supplies. It has, in fact, been estimated that during the 1960s competition from foreign oil would have necessitated the closure of between a third and a half of domestic oil production. Such competition would have particularly affected production from the coastal or near-coastal fields, as these lacked the protection given to the interior fields by the cost of transporting foreign oil to the inland regions of the very extensive American land mass. This estimate was based on an evaluation of the market opportunities for imported oil, which during this period could have been landed at U.S. east- and Gulf-coast refineries at a price of $1·50 per barrel. In the light of the production and transport costs of providing U.S.-produced oil to these refineries, indigenous oil production levels of over 500 million tons per annum were achieved only with a well-head price level which was more than $3 per barrel. If the refiners had been allowed to obtain their crude oil requirements at as little as half this figure one can see that the production of much expensive domestic oil would have been eliminated.

This cost of the mandatory oil quotas policy obviously kept up the price of oil to the consumer, and in end-uses where there were no possibilities of substitution by other sorts of energy the consumer had no alternative but to pay the additional cost. In many end-uses, however, other sources of energy were substituted for oil as the oil import policy kept prices up. Natural gas – whose price in the U.S.A. was, in contrast with oil, kept down by government intervention – benefited, particularly in home heating and similar markets. In addition, American coal also enjoyed additional markets at the expense of foreign oil. It was able to increase its markets in thermal electricity generating plants, not only in the areas of coal production themselves, but also in the east coast megalopolis stretching from Boston to Washington. This would otherwise have provided outlets in bulk for cheap foreign oil.

Internally, therefore, a large number of powerful pressure groups were well pleased with the mandatory oil quotas. This was also the case with those oil refiners in the U.S. who were eligible for 'tickets' giving them the right to import specified quantities of foreign crude oil. These rights, indeed, became negotiable currency. With a price differential between domestic and foreign oil of about $1·50 per barrel it was obviously more profitable for a refiner to pay, say, $1 for a 'ticket' which gave it the right to import a barrel of foreign crude. The overall cost to the refinery was still 50 cents less than if it had used indigenous crude oil.

Thus a single act of government (which remained effective for over a decade, as amendments to the quota system between 1960 and 1972 only relaxed the restrictions to some degree) ensured that the U.S.A. remained the world's largest producer of petroleum throughout the 1960s. Without the oil import quota system it would certainly have been overtaken much sooner by the U.S.S.R., probably by Venezuela and possibly by one or more of the Middle Eastern producing nations as well before 1970. The system also encouraged the continued search for new petroleum resources within the U.S.A. by increasing the profitability of the oil industry and so provided the incentive for more exploration and development. The discovery of the Alaskan oilfields can indeed be said to be mainly the result of import quotas. Oil discovered in Alaska would, of course, enjoy free access to the protected U.S. market, so there was every incentive to find and to develop it, in spite of the state's physical separation from the rest of the country, involving high transport costs, and the difficult environmental conditions there for oil exploitation.

The decision of the government to restrict oil imports, however, also produced a situation in which U.S. oil reserves were run down faster than would have been the case if unlimited foreign oil had been allowed into the country. Excluding the discovery and development of the Alaskan potential, the U.S. domestic oil industry had to struggle for the twenty years from 1959 to maintain reserves sufficient to sustain the higher production levels required to meet the growing demand for oil. It was therefore argued that, if reliance on foreign oil was considered to be a security risk, then the best policy to pursue was one in which imports were encouraged in the short term; in a period, that is, in which they remained readily available. This would have conserved U.S. domestic resources for the time when the potential risk from dependence on imports became an actual one; as, indeed, in the period since 1973. This line of argument was never effectively answered, suggesting that the protection of domestic energy producers' profits rather than concern for the security of overseas supplies was the more important element in American oil policy-making in the period from 1959 to 1970.

After 1970, however, an energy shortage developed in the U.S.A., a result in large part of the control exercised over natural gas prices by the Federal government. These controls inhibited the discovery of enough reserves to sustain an unexpectedly high rate of growth in demand which, in turn, was largely a function of the increasing use of electricity, the production of which is very energy intensive (two-thirds of the heat value of the fuel used to make the electricity is 'lost' in the process). The threat this posed to the smooth functioning of the U.S. economic and social systems necessitated a reappraisal of U.S. policy towards oil imports. It was no longer appropriate to keep imported oil out of the country. Thus, in mid-1973 import controls were lifted, except for a small remaining tariff designed to give protection to some domestic producers and to the U.S. oil-refining industry. However, given the strong upward movement in world oil prices after 1971, the degree of protection that was really needed by that time to keep the U.S. oil producers in business was small, and even this need disappeared after the large increase in international oil prices between October 1973 and February 1974. Thereafter, the more important question was how to raise the profitability of the domestic oil producers, in a period of government controls over oil prices, so that their interest in finding and producing more oil in the U.S.A. would be stimulated – not only from the newly-discovered oil in the remote Alaskan fields, but also by enhancing the rate of recovery of oil from the high-cost, onshore oil-bearing structures from which, on average, less than 30 per cent of the oil in place had been recovered, and from possible oil reserves lying offshore on the Atlantic and Pacific continental shelves of the United States.

A continuing dispute, however, between the oil industry and the government over the degree to which U.S. oil prices should be allowed to rise to the very much higher post-1974 world oil prices inhibited the development of these new and additional U.S. resources to any large extent for the rest of the decade. Measures to eliminate this adverse effect of price controls were at last taken in 1979–80 and, thereafter, the decline in domestic oil production was halted as the financial and other inducements which were offered to the industry, first by the Carter administration, and then by President Reagan, began to take effect. The latter was determined, even more than his predecessor, to minimize the country's dependence on imported energy and he thus implemented his commitment in full to remove government controls – and particularly price controls – on U.S. oil industry enterprise. Since 1981 production has steadily increased – albeit relatively slowly – in spite of the fall in oil use over the same period. The overall result has, of course, been a dramatic fall in U.S. oil imports which in 1984 were less than 40 per cent of their peak levels of the late 1970s.

The maximization of domestic oil production became particularly important in the light of the experience of the Arab oil-producing nations' embargo on oil supplies to the U.S.A. in 1973-4 as a means of bringing pressure to bear to get the U.S. to change its policy in the Middle East, and then in the second half of the 1970s as a result of the adverse effect on the dollar of very costly imports of oil. Increased indigenous production has initially been mainly from the large Alaskan discoveries, the development of which finally proceeded after a delay of several years for environmental reasons. More conventional oil from the 'Lower 48' states is, however, now also being produced under the stimulus of higher prices. This has been achieved partly by technical developments, such as the use of new methods of recovery designed to produce quantities of oil which cannot otherwise be extracted from certain oil-bearing formations because of particular geological and other physical conditions, and partly as a result of the encouragement which has been given to find new fields – particularly at greater depths – in old oil-producing areas. Similarly, decisions were also taken to give substantial government support to the extraction of oil from the oil shales which are present in effectively limitless amounts in several states of the Union. Their development was expected to enable oil to be produced at a cost below the price of oil in international trade in the mid-1980s and a fairly rapid development of this new industry was anticipated in response to the emphatic wish by the United States to be as self-sufficient as possible in oil – the same motivation, in fact, which originally lay behind the introduction of the mandatory quotas on oil imports in 1959.

Thus, in overall terms, the short-term losses for the American economy which arose from the higher energy prices consumers had to pay between 1959 and 1972, as a consequence of the introduction of oil import quotas, are now being offset by the benefits of the technological and other developments in energy production that have been encouraged since the late 1970s. These promise, in the longer term, to help the U.S.A. to maintain its economic lead over the rest of the world.

It is particularly ironic that the imposition of quotas and other elements in a generally autarkic energy policy have produced the hitherto unforeseen possibility of the U.S. redeveloping an oil export potential. This possibility is based on oil from Alaska, whose oil resources could, indeed, best be exploited not to serve the U.S. market, but markets in other parts of the world more accessible for oil tankers from Alaska. Alaskan oil exports – particularly to Japan and other parts of the Far East – could, within a decade, help to offset the remaining need for net oil imports by the U.S. from other parts of the world. However, such exports were forbidden by President Carter in order to eliminate the risk of their reducing the degree of self-sufficiency in the short term. The present Reagan administration has recently recommended that this policy should be changed if more oil is discovered in Alaska. This oil could not be marketed in California and other west coast states, and it is not possible to reduce the high costs of transporting it to the U.S. east of the Rockies. Thus, Reagan argues, U.S. economic interests could best be served by allowing the companies concerned to sell such oil abroad. This would be particularly true if, as indeed is already happening, U.S. oil use declines – not simply below the levels which were expected, but below levels previously achieved – as a result of conservation in use (particularly in the transport sector), and its substitution by other sources of energy elsewhere in the economy. Should these circumstances continue, then the United States may yet again become a not unimportant oil-exporting country.

But if the results of the mandatory quotas and autarky in respect of oil have not been entirely disadvantageous for the U.S.A. itself, the same cannot be said for other countries of the world. These, of course, include the ones which produce oil for export. It has been shown earlier in this chapter how many U.S. oil companies eagerly sought concessions to search for oil abroad and, in many cases, quickly achieved success, thereby raising the hopes of the host countries for important new sources of revenue and employment and for other economic advantages. These companies had entered into these expensive commitments in order to find, produce and export oil back to the U.S.A. The mandatory quotas thwarted these plans – even more effectively than would perhaps at first appear in the light of the way in which the quota system worked.

The quotas were not allocated on a percentage basis to all who requested them, but were given to those companies which were 'traditional' importers of oil and oil products (that is, those companies which had been importing oil during the immediate post-war period). Thus, new-comers to the international oil business were virtually excluded from selling oil to the U.S. unless, of course, they were prepared to purchase a 'ticket to import' from one of the 'traditional' importers. The latter thus stood to make profits much more easily out of selling their rights to import than out of continuing to trade in foreign oil. This happened to some extent but, by and large, the American companies which had discovered oil abroad with the intention of selling it to the U.S.A. were forced either to sit tight on their discoveries or to seek markets elsewhere. It has been estimated, as previously indicated, that the quotas led to some 150 million tons more oil per annum being produced in the U.S.A. than would otherwise have been the case, and so overseas producers 'lost' an equivalent market. In addition, higher oil prices in the U.S.A. in the 1960s, as a direct result of the import quotas, could well have reduced oil consumption by 100 million tons per annum compared with the size of the market if cheaper foreign oil had been available. Thus, in total, the major oil-producing nations were denied the opportunity to produce each year some 250 million tons of oil for the U.S. market. Even at pre-1973 oil prices this amount of oil produced for export would have generated revenues for the governments concerned of at least $2,500 million, together with other beneficial effects arising out of the production and export of this additional oil.

Yet even this is not the total measure of the effects of the U.S. decision to introduce oil import quotas. The decision also meant, as indicated above, that some of the companies concerned had to find markets for their oil in other parts of the world, in order to obtain some income from sales to offset their exploration and development expenditure. This, in turn, proved to be the single most important factor in upsetting the price equilibrium which had been established, outside North America and the communist countries, by the orderly marketing arrangements of the seven large international companies in the early 1950s. Their control over the great majority of producing operations and marketing outlets had hitherto meant that prices could be kept well above the cost of oil production.

In order to break into this orderly market the new companies, with crude oil available, at first offered it to independent refiners at prices below those formerly charged by the international majors. Then they went further and started to build refineries themselves in order to enhance their marketing opportunities, both geographically and in terms of the variety of products which they were able to offer. The international companies reacted to this challenge to their control over the system by reducing their prices where necessary to keep business, and, at a later stage, they were even obliged to reduce prices to their own subsidiaries in order to enable the latter to compete and remain profitable in the local markets with which they were concerned, and which were becoming subjected to competition from the independents.

Thus, the oil that became available on the world market as a result of import quotas in the U.S.A. acted as the catalyst to upset the earlier post-war norm of orderly marketing. As prices in the market place fell away so this was reflected back in efforts by the companies to contain and even reduce their payments on each barrel of oil to the host governments, which as a result saw their revenues and economic livelihood being threatened. Out of this, as we saw in Chapter 1, emerged O.P.E.C. and a subsequent battle with the oil companies to try to ensure that the producing countries did not lose revenues by virtue of the price weakness which was developing. The collective strength of O.P.E.C. was

45

eventually sufficient to inhibit further royalty and other tax reductions which the companies would otherwise have imposed. On the other hand, the market weakness caused by U.S. policy certainly eliminated the ability of the countries concerned to secure higher returns from the level of oil exports that might otherwise have been achieved, had the action of the U.S. government in limiting oil imports not upset the delicate equilibrium of the world market.

U.S. government action thus rebounded unfavourably as far as the oil-producing nations in general were concerned. In particular, however, it proved seriously disadvantageous to Venezuela, which, in the face of the increasing production of lower-cost oil from the Middle East, had gradually lost its markets in Western Europe and so became dependent largely on United States markets. U.S. import quotas virtually eliminated Venezuela's only significant growth market overnight, and the oil economy of the country suffered for over a decade. Successive Venezuelan governments thus made repeated efforts to secure some form of preferential treatment based on the country's particular geo-political relationships with the U.S.A. and within the framework of western hemisphere treaty arrangements. In economic terms Venezuela considered the U.S.A. to be its 'natural' market. Though such a concept had little economic validity in a situation of widely varying production costs at different locations, and of low and falling ocean freight rates which eliminated the competitive transport cost advantage of the nearer supplier, the Venezuelans continued to make use of the concept and the idea thus achieved reality in political terms. In geo-political terms, moreover, Venezuela occupied the key position in the Caribbean area following the defection of Cuba to the communist alliance. Venezuela used these facts, coupled with a constant reiteration of the U.S.A.'s original concept of 'hemispheric solidarity', to try to persuade the U.S.A. to make an exception for Venezuelan (and other Latin American) oil, similar to that granted to Canada and Mexico. Venezuela even stressed that it would not take undue advantage of its exemption from the quota system and that it would be prepared to negotiate reasonable marketing opportunities, in the same way that Canada had voluntarily limited its exports to the U.S.A. at the time. Alternatively Venezuela argued that its oil (or Latin American oil in general) should be given a privileged position within the quota system, possibly by guaranteeing it a specified share of the total and of the annual incremental tonnage permitted under the system.

These were powerful arguments, and the U.S. administration was several times on the brink of accepting Venezuela's pleas in one way or another. But each time it drew back from the brink and refused special treatment to Venezuela, except over issues not affecting U.S. policy in general terms – such as freeing the Puerto Rican market for petro-chemical feedstocks from the overall quota arrangements. This was a market which Venezuela was particularly well placed to serve and for which it also had appropriate types of crude oil and products readily available, and thus the concession was a helpful, though small, positive response to Venezuela's pleas. The U.S. also freed imports of residual fuel oil in general. Though this step happened to be to Venezuela's advantage, it was taken essentially for internal U.S. political reasons; namely, to avoid too high industrial and residential oil prices in the New England and New York/Washington areas. The U.S.A.'s refusal to take any further specific action to help Venezuela – a refusal seen in Latin America as another example of the way in which the U.S.A. fails in its duties on the continent, to the support of which it gives a great deal of lip service – was based, in public, on a declared anathema to the concept of government interference in the right of private oil importers to buy their oil where they choose. In reality, the refusal was related to an unwillingness by the United States to upset the oil-producing nations in the politically unreliable Middle East. The public reasoning rang very hollow indeed, in the light of the fact that the imposition of oil import quotas was in itself a much greater interference with the right of private enterprise to choose its own supply sources. The diplomatic reasoning, on the other hand, was not without some foundation, although Vene-zuela said that its fellow member countries of O.P.E.C. indicated that they would not have been very upset by a decision of the U.S.A. to accede to Venezuela's pleas. Their 'indications' might have been given, however, in the pretty certain knowledge that their sincerity over the matter would not be put to the test by the U.S.A. Thus, though many other factors combined to produce difficulties for the Venezuelan oil industry in the 1960s (these are considered in Chapter 4), the decision of the U.S.A. to restrict its imports of overseas oil was certainly one of the most important.

But if Venezuela's loss as a result of the U.S. oil import quota system, in terms of government revenues, jobs and new expenditure by overseas oil companies in the country, can be estimated at many hundreds of millions of dollars each year between 1959 and 1971, the same quota system produced economic gains for other nations in other parts of the world – notably Western Europe and Japan, whose energy consumers steadily benefited from the reduced oil prices arising out of the world 'surplus' of oil created by the U.S. quota system. Though this 'help' to

Europe and Japan on the part of the U.S.A. was quite inadvertent, it seems not impossible that its importance to the economies of Western Europe was as great as the effects of the Marshall Plan for European recovery. Moreover, while the Marshall Plan had a time limit and depended for its funding on a U.S. Congress which examined its expenditure very critically each year, the beneficial effects to Western Europe of the low oil prices went on for year after year without the inquisition of a Congress which, for other reasons, was unlikely to do anything to end the oil quota system from the effects of which on the international oil market European and other countries were benefiting. The U.S.A. had had energy cost advantages over the rest of the manufacturing world from at least the beginning of this century. This advantage was eliminated – or all but eliminated – after 1959 by the policy of the United States which created an availability of very low-cost oil to Europe and Japan. One can perhaps speculate that this development provided one of the reasons why the U.S. found the economic going so difficult in the 1960s and early 1970s when it only managed to achieve economic growth rates significantly below those of most other industrialized nations.

Thus, though the oil industry can correctly be considered an American invention and one, moreover, which has generally remained American dominated, it can also be argued that some parts of the rest of the world learned not only how to live with this American industry, but also how to take such economic advantage of it that their gains from the oil industry over the period of nearly fifteen years from 1959 to 1973 clearly exceeded the benefits to the U.S.A. In this period both oil-producing and oil-consuming countries increasingly 'called the tune' in the system of the international oil industry – and the U.S. companies involved jumped to it when called. They then had the U.S. taxpayer to bail them out when the going got too tough. The belief that 'U.S. domination of international oil' was part of an 'unfair' international economic system thus began to have a somewhat ironic aspect to it in the 1960s. At the beginning of the 1970s the situation in this respect seemed to be getting even more ironic – as it then appeared that the U.S. would have to compete on the world market for increasingly large supplies of oil which, at least temporarily, the United States could not produce for itself. Moreover, even after having been successful in buying the oil, it still had to face difficulties in transporting it to the U.S.A. and in refining it along the east coast – given the success of the environmentalists in preventing the construction of terminals for mammoth tankers and of new refineries to run on the foreign crude oil in this part of the United States, in response to which special costly arrangements had to be made to tranship imported oil through terminals in the Caribbean.

From 1973, however, America had an opportunity to reassert its influence in oil matters. The Arab–Israeli war of that year led to an embargo on oil to the United States. This was followed by the traumatic rise in international oil prices as the oil-producing countries, working effectively together, exercised their power. The United States stood to benefit from the radically changed situation. First, it gained initially from the greatly enhanced profitability of the international oil companies in the aftermath of the oil-supply crisis. Most of these companies are, as shown in Chapter 1, American owned and controlled and so their increased flow of profits helped the U.S. balance of trade position. Second, given the limited extent at the time to which the U.S. economy depended on oil imports, it was able to divorce its energy pricing system from that set by the price of international oil. In doing so it was able to provide its consumers with a very significant energy cost advantage (of 50 per cent or more) over their rivals in other parts of the industrialized world. This is particularly true, of course, in respect of Western Europe and Japan which, as will be shown in Chapters 5 and 6, had allowed their economies to become very largely dependent on foreign oil. In the radically changed oil power struggle of the 1970s, they found themselves in difficulties over both the supply and the price of oil.

Thus, the United States should have been able to take significant advantage of the new situation, particularly in the light of its close relationship with and influence over Saudi Arabia, the most important oil-exporting country. Unfortunately for the United States, however, it was not successful in increasing its production of domestic oil and other sources of energy quickly enough, in large part because of the failure of the government and the oil companies to agree on price, tax and other issues. Neither was it as successful as other countries in constraining the demand for oil in the aftermath of the 1973-4 crisis. As a result its need for oil imports continued to grow quickly so that the increasing cost of foreign oil more than offset the benefits which the U.S.A. gained from the post-1973 change in the international oil system. The speed at which its imports of oil could be reduced, and by how much, thus became a question of critical importance for the United States.

We have previously described the recent major steps taken to secure such a reduction and have noted that a degree of success has already been achieved – by means of demand constraints and supply encouragement. U.S. oil imports were back, in 1984, to less than 40 per cent of their level of the late 1970s. There remains some degree of pessimism concerning the continuity of the successes achieved – largely because of a continued lack of trust between large sectors of the population of the United States, on the one hand, and the oil companies, on the other, so that action which inhibits the willingness of the latter to continue to maximize oil output in the United States is always a strong possibility. This pessimism will become an even stronger possibility if the second Reagan administration, which is thought to have favoured its 'friends' in the industry, is followed by a Democratic government. Meanwhile, as we shall see later, the oil companies have lost much of the control they hitherto exercised over the international oil system and this development, too, given the 'American-ness' of the oil companies, constitutes a deterioration for the United States in respect of its role in the world of oil power.

3. Soviet Oil Development

Amongst Soviet claims to have led the world in the field of technical developments is one which attributes the world's first oil well to the Soviet Union. This claim challenges the more generally accepted view that the modern oil industry began in the United States in 1859 (see p. 25), but adjudication in this dispute is not relevant to the task in hand. The fact that the Soviet claim can be made at all does, however, indicate the early date of oil resource exploitation in Russia. The expansion of this early initial exploitation of Russian oil into a Russian oil industry which was already of world significance by the late nineteenth century was a function of European trading interests. They sought and obtained concessions in Russia in order to provide Europe with an alternative to the United States as a source of oil supplies, and thus break the near monopoly which Rockefeller's Standard Oil Company had achieved in Europe with American oil products. The shallowness of the oil deposits and other attractive geological conditions in the vicinity of Baku in the Caucasus, together with the completion in the 1880s of good transport facilities to the Black Sea coast, encouraged the growth of Russian oil production. At the same time the proximity of the oil to potential markets, particularly those in Europe, compared with the alternative supply points in the U.S.A., made these early Russian ventures financially interesting. A little later the rapidly emerging Royal Dutch Shell Group also used its Russian oil production to break the monopoly of Standard Oil east of Suez. Success in that part of the world, however, depended on the company's ability to deliver the oil in bulk, rather than in tins or drums, whose shipment through the Suez Canal necessitated expensive safety precautions. Success in this respect was achieved in 1892, with the entry into service of an ocean-going tanker which was given safety clearance by the Canal authorities. As a result Russia's oil development was given an important boost and in the period before the First World War it was second only to the U.S.A. in its total oil production. In fact, for a few years Russia was even ahead of the U.S.A. in its annual production of oil. Through to the outbreak of the First World War and to Russia's involvement in it, physical conditions for oil production in the Baku region remained attractive, the markets of Western Europe continued to grow, and foreign capital continued to be drawn to this important producing area. Thus, the prospects were set fair for the rapid and continuing development of the country's oil resources under the oil concession system which the Czarist régime was content to accept, in return for the payment of royalties by the oil companies concerned. This typical nineteenth- and early twentieth-century approach to the development of a country's mineral resources by foreign companies, mainly from the United States and the United Kingdom, was, however, first upset by the involvement of Russia in the First World War and was then terminated abruptly by the Soviet Revolution in 1917.

As a result of the revolution the oil industry in Russia, like every other sector of the economy, passed into the ownership of the state. In the chaos of the immediate post-revolutionary period, Soviet oil production declined markedly and its export trade in the commodity virtually disappeared. However, as the Soviet authorities gradually gained effective control over the country's economic life, the Baku oilfields were reactivated under the stimuli of a growing domestic need for oil and the possibility of reviving exports, whereby the Soviet Union could earn some of the hard foreign currency which it so desperately needed for its economic development. Production from Baku, therefore, gradually recovered from the low levels of the early post-revolutionary years and by the late 1920s the state-owned and run oil industry of the Soviet Union was producing more oil per annum than had been produced in any year in the pre-revolutionary period. This success, however, must be put into perspective, for it took place at a time when world oil production was increasing even more rapidly under the stimulus of a big growth in demand generated by the initial development of the motor-car as a mass consumption product (notably in the United States but also, though to a lesser degree, in the wealthier countries of Western Europe) and by the increasing use of oil in the industries of North America and of Western Europe. The Soviet Union's production efforts did not in this period keep Soviet oil production moving ahead as rapidly as the overall rate of increase in an increasingly oil-conscious world. The country did, nevertheless, produce enough oil to meet requirements in the U.S.S.R. itself but, it should be noted, these requirements were limited by the deliberate choice of the U.S.S.R.'s policy-makers not to expand its motor industry or to encourage the use of motor transport. Elsewhere in the Soviet Union the growing number of factories, springing up under forced industrialization, could be fuelled by coal and lignite available in closer proximity to the main industrial areas of Moscow and Leningrad, or from the major hydro-electric schemes being developed on some of the country's important rivers.

Soviet oil was, however, increasingly successful in foreign markets, and to encourage this development, which was highly desirable – indeed, even essential – from the point of view of the U.S.S.R.'s need to earn foreign exchange to use in its development plans, the Soviet government invested in tankers, oil depots and distribution facilities in many countries in Western Europe including the U.K., where products were sold under the trade name, R.O.P. – Russian Oil Products. As a result of a relatively aggressive sales policy in the chosen markets of Western Europe, the Soviet Union achieved about a 15 per cent share of the total market available. Although this share fluctuated in the 1930s, largely because increasing demand in the Soviet Union itself created pressure from time to time on the country's limited producing facilities, this general level of penetration of Russian oil into Western European markets still existed at the outbreak of the Second World War in 1939, and so made the Soviet Union second only in importance to the United States in supplying oil to Western Europe.

During the Second World War most of the Soviet Union's existing oil-producing facilities were overrun by German armed forces. Thus the Soviet Union had to seek oil in other parts of the country, over large parts of which there were sedimentary basins which were potentially petroliferous (see Map 1, p. 54). However, wartime difficulties and other conditions made exploration for oil and the development of production almost impossible and, for most of the war, much of the Soviet Union's war effort depended upon the success of the convoys of oil tankers and other ships moving around the coast of northern Europe, to the north Russian ports of Archangel and Murmansk, and on the overland movement of oil by rail from Iran to the southern parts of the U.S.S.R.

At the end of the war the Soviet Union initiated the massive reconstruction of its economy and one of its earliest efforts was to secure the rapid expansion of the country's oil and gas resources. Wartime oil exploration had indicated the existence of large fields in the area between the Volga and the Urals and these were now rapidly developed, so that they could immediately supplement the output of the rehabilitated fields of the Baku area. Within a few years, however, their size enabled them to dominate the pattern of Soviet oil output, and the new area of production became known as the 'Second Baku'. Although the Soviet economy was still essentially based upon the use of coal and other solid fuels, and although preference in capital investment programmes tended to be given to the development and expansion of hydro-electric resources, the

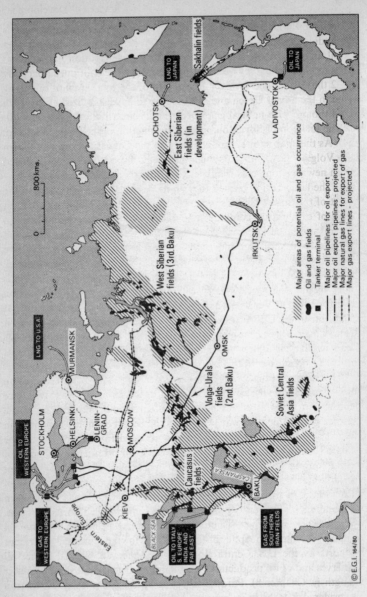

Map 1. Soviet oil and gas production and export facilities.

share of oil and gas in the total energy requirements of the country's expanding economy started to increase. Since the early 1950s increases in oil and gas production and an increasing contribution of these two fuels to the total energy needs of the economy have continued apace. By 1982 they accounted for over 70 per cent of the total amount of energy used in the Soviet Union and now appear to have stabilized at this high level though with natural gas increasing its contribution relative to oil.

The location of Soviet oil and gas production is shown in Map 1 (p. 54). As this map shows, production from the Caucasus region and from the Volga–Urals field is now supplemented by oil (and gas) produced from new developments in Soviet Central Asia and in western Siberia – with the latter region termed the 'Third Baku', and currently providing most of the Soviet Union's incremental production. Moreover, the fast pace of exploration continues and the search has been extended into central and eastern Siberia, where the potential for oil and gas production is considered very high indeed, and where important discoveries have already been made. The old and new areas have together yielded well over 600 million tons of oil per year since 1980 and the Soviet Union is now by far the world's largest oil producer. Soviet oil production overtook that of Venezuela as long ago as 1961 to make it second only at that time to the U.S.A. It exceeded the latter's production in 1976 and it is now over 20 per cent greater. It is now clear that it will maintain this lead indefinitely unless Alaska or other frontier areas in the U.S.A. prove to be more prolific than is currently expected.

This success of the Soviet Union in developing its own oil resources on such a large scale has reduced, if not entirely eliminated, what was once considered to be a serious potential threat to the Middle East in the earlier part of the post-war period. As the Soviet economy expanded in the 1950s many observers argued that its need for oil would lead the Soviet Union to cast covetous eyes on the oil production capacity of neighbouring countries to the south, particularly Iran and adjacent Middle Eastern countries like Iraq and the oil-exporting sheikdoms of the Gulf. The great successes which the Soviet Union has achieved in expanding its domestic production have undermined this argument. The danger, moreover, is unlikely to reassert itself, at least on oil supply grounds, in spite of speculation to the contrary in recent reports by the U.S. Central Intelligence Agency of limited Soviet oil reserves and of oil production potential (the issue is discussed at greater length in Chapter 8). Paradoxically, however, in the early 1970s Iran considered it to be in its own national interests to reach a commercial agreement with the Soviet Union whereby Iran was to supply adjacent areas of the U.S.S.R. with some of the associated natural gas which it produced from its southern fields. This gas, produced as a necessary by-product of oil production, had no markets in Iran – except local industrial ones that have been slow to develop – and a large diameter natural gas line, the capacity of which was increased by over 50 per cent by 1976, was constructed from the southern Iranian fields into the Soviet Union, where it linked up with the pre-existing Soviet natural gas pipeline system. The agreement between Iran and the Soviet Union was certainly in the national interest of the former – with no other immediately foreseeable large market for its gas production – and was important for the latter in that it represented a new willingness to import energy in order to save the considerable amounts of capital which would otherwise be required in order to make indigenous energy available to the areas concerned. At the same time it enabled the Soviet Union to make equal amounts of gas available from more favourably located Soviet fields for export to Western Europe. Thus, the Soviet Union was effectively able to act as the 'middleman' in the initial development of Iranian natural gas exports to Europe. Without this arrangement with the Soviet Union, Iran could not have marketed the gas involved. The agreement can thus be interpreted very satisfactorily in economic terms. Its political overtones were minimal, though these have become more important following the revolution in Iran and as a result of which Iran closed the pipeline and cut off supplies to the U.S.S.R. By the time this happened, however, the Soviet Union had further expanded its own natural gas production and its export potential so that the cessation of supplies from Iran created only short-term supply problems in the areas of the Soviet Union where the Iranian gas had been used: and neither did it lead to other than a temporary fall in the amount of natural gas which the Soviet Union sent to Western Europe.

Iran has now reactivated the deal while a similar agreement, whereby the Soviet Union imports smaller quantities of gas from fields just over the border of Afghanistan, has recently been expanded. These have been the only significant ventures by the Soviet Union into an oil and gas import policy. Otherwise the basis of its oil and gas policy has always been, and appears to remain, one which has the aim of achieving and maintaining self-sufficiency, no matter what the cost. Thus, although certain parts of its territory – particularly the far east and the far north – could probably have imported their energy requirements at a lower cost to the Soviet economy than that involved in making domestic energy available to these areas, there has been no response by the Soviet Union to offers from time to time in this direction by various oil companies and other traders from the western world. A national pipeline system and local distribution networks have been built to ensure that most parts of the Soviet Union, including all those with a significant degree of industrialization, can be supplied from the country's oil- and gas-producing areas. Expansion of production is being continually attempted in the light of the planned future growth of energy consumption and the arguments on the danger of relying on other countries for the supply of commodities which are essential for national development remain completely accepted. This is particularly important given the frequent upsets in the international oil market over the last decade and it now seems very unlikely that there will be any fundamental change in the Soviet policy of seeking complete autarky in the energy sector of its economy as a prime strategic objective.

Soviet interests in international trade in oil have, in fact, been in the opposite direction since the mid-1950s when the U.S.S.R. reverted to its pre-war policies of seeking export markets for part of its production. It has, more recently, also sought export markets for natural gas. There seem to have been two distinct motives behind the Soviet Union's post-war search for oil and gas markets abroad, the first wholly economic, and the second economic to some degree, but with political overtones. In respect of the first motivation it has taken opportunities to export its oil to countries where the oil could earn 'hard' foreign currencies, or where oil exports could be used in barter-type arrangements for securing goods the U.S.S.R. requires in its development programmes. For this type of trade, crude oil and petroleum products are almost ideal exports for the Soviet Union in that the purchasers in the western industrial world know exactly what they are getting, because there are internationally agreed specifications which have to be met. Oil exports are thus less likely to be technically or otherwise suspect than other products of the Soviet Union's relatively recent industrialization, given that many of its manufactured goods have not been subjected to the rigours of western-style consumer evaluation.

Thus, Soviet oil exports constitute desired and acceptable goods, competing in a perfectly straightforward way with similar products from elsewhere as far as Western European countries and Japan are concerned. This view was, however, only gradually accepted in the West. Its final acceptance was in large part a function of the retreat from the Cold War attitudes which still prevailed when the Soviet Union initiated its first post-World War Two oil export campaign in the mid-1950s. Its attempts to sell oil in Western Europe at that time were interpreted in political terms by the U.S. State Department and some of the major

American oil companies. Western European nations were thus 'warned' of the great economic risks involved in relying on an energy source which, it was argued, could be turned off at the whim of the U.S.S.R. policymakers. The opponents of Soviet oil imports made more than full use of the one occasion on which the U.S.S.R. had declined to honour an oil supply agreement as a result of changed political circumstances. However, as this was the special case of a Soviet refusal to go on supplying Israel in 1957, after it had been declared an aggressor nation by the United Nations for its invasion of Egypt, it was hardly acceptable as a valid precedent. Indeed, the European nations which became interested in Russian oil were able to turn such arguments back on their instigators when they showed just how dependent the same American companies were on Middle Eastern Arab countries for their oil supplies, in a situation in which these countries argued that their willingness to continue supplying oil depended on the recipients' willingness not to help Israel in its battle against the Arab nations. Britain and France, of course, suffered from the political whim of the Arab oil-exporting countries in this respect in the months after their collusion with Israel in the 1956–7 Suez crisis, as did the Netherlands specifically, and Western Europe generally, in the aftermath of the 1973 conflict in the Middle East.

In that there are underlying political factors which involve almost the whole of the world oil industry in one way or another, most European nations were somewhat sceptical, to say the least, of American efforts to 'knock' Russian oil in this way. Thus, the arguments were largely ignored and Soviet oil gradually found markets in the countries of Western Europe, as both state entities and private refiners and marketers found some Soviet offers acceptable. Of course, as a new entrant to the market at a time when most outlets were controlled by the major international companies, the U.S.S.R. could only succeed in achieving its export targets by offering oil at cut prices. But there is no evidence to suggest that prices were ever cut more than the commercial expertise of the Russians indicated that they should be, in order, as some people suggested, to enable the Soviet Union to undermine the economies of Western Europe! On the contrary, there is evidence available which points to a Soviet unwillingness to sell its oil at the very depressed prices which obtained from time to time in some Western European markets in the era of declining real oil prices in the late 1950s and the 1960s; as, for example, when the Soviet oil export agency declined to bid for oil sales in Switzerland in the early 1960s, at a time when excessive volumes of oil were on offer from a very large number of companies for the supply of oil to the relatively few outlets in Switzerland which were not tied to individual major oil companies' supply patterns; or more recently in 1985 when the Soviet Union appeared to limit its oil products' sales on the Rotterdam spot market because prices were considered to be too low.

Supporters of the view that Soviet oil was being 'dumped' in Western Europe, and/or sold to secure some political or strategic advantage usually pointed to the wide discrepancy between the prices of Soviet oil in Western Europe and the much higher prices which were charged to Eastern European countries. What such observers were in effect pointing out, however, was a good example of the great disadvantage of being a captive customer, in a time of 'over-supply'. The high oil prices paid by the countries of east Europe, which lacked the freedom to buy oil from the west both for political reasons and because their long-term bilateral trading treaties with the U.S.S.R. inhibited their rights to negotiate their oil supplies from elsewhere, were not the only countries in this position in the world of oil at that time. It also applied in most Western European countries and in most nations of the Third World. In these countries even higher prices for supplies had to be paid by local refining and marketing companies of the international oil corporations, in a situation in which the latter were simultaneously selling crude oil and products much more cheaply to non-associated companies and to state entities, for whose business they had to compete.

To a large degree, however, the particular motivations which lay behind the Soviet oil export drive in Western Europe did little more than provide a subject for academic debate; a topic for discussion in international bodies such as NATO and the Council of Europe, under pressure from the member countries with interests in international oil companies; and a convenient explanation for some oil companies in their efforts to explain away their own higher prices in markets where Soviet oil was available at much lower prices. While debate and discussion were raging many individual countries tested Soviet oil, found its specifications to be acceptable, and proceeded to take advantage of it in the interests of their own economic development.

Sweden was in the forefront of the partial switch to Soviet oil, the importation of which it quite deliberately built up to some 15 per cent of its total oil import requirements. Significantly it was Sweden – unencumbered by membership of NATO and the pressures that were brought to bear on the members of the Organization by its oil company owning nations – which first indicated that the usual security-of-supply arguments against Soviet oil could, in fact, be reversed. Sweden pointed out that its increasingly oil-based economy depended upon political stability in the Middle East and the willingness of countries there to continue to supply oil. Such a situation not only implied security dangers; it had already given rise to difficulties both in the early 1950s, as a result of Iranian nationalization of its oil, and in the later 1950s, as a result of the Suez crisis. In such circumstances oil from the Soviet Union – imported as part of a bilateral trading arrangement in which Russia obtained goods that it really needed from Sweden and from the sale of which the Swedish economy benefited – reduced the security risk. Sweden also pointed out that transport of Soviet oil to Sweden was less liable to interruption than that of oil from other sources, particularly with the completion of the pipeline from Soviet oilfields to the Baltic Sea (see Map 1). Only in the event of a major war involving the great powers would Soviet oil supplies to Sweden be at risk, and in such circumstances it was unlikely that many nations of the world would be very worried for very long about the specific problem of their oil imports.

The significance of Sweden's early success in its oil dealings with the U.S.S.R. was not lost on other European nations: one by one they accepted the validity of the economic arguments and discounted the alleged security-of-supply risks as originally argued officially by NATO under extreme American pressure. Even West Germany became willing to negotiate for large supplies of Soviet oil in exchange for Russian markets for its iron and steel production and for export openings for other manufactured goods. France and Belgium similarly encouraged Soviet oil trade – though neither to the same degree as Italy, which deliberately set out to take the maximum possible economic advantage of large-scale Soviet oil and gas availability.

This attitude stemmed from Italy's evaluation of oil as a source of energy which would help to eliminate the country's long-standing re-source and locational disadvantages in the field of fuel and power, in comparison with the more favourably blessed countries of north-west Europe. Low-cost imported oil promised much in this direction, but Italy soon found that access to such oil was largely denied to her through the organization of the international petroleum industry which, working within the framework of a system of posted prices, kept prices up. Italy's immediate reaction was to give more or less *carte blanche* to E.N.I., the state oil entity, to seek oil abroad. Its first efforts lay in finding and developing oil resources abroad, but it quickly became obvious that success in this direction would take too long to achieve and, in the meantime, the Soviet Union had appeared on the scene as a large-scale supplier of oil at prices well below those that Italy could obtain elsewhere. Soviet oil thus served both the political and commercial interests of

E.N.I. and the economic interests of the nation and, from small beginnings in the mid-1950s, E.N.I. steadily built up its dependence on Soviet supplies. To encourage this development the Soviet Union responded by offering preferential outlets for Italian goods, and the two nations eventually signed a bilateral agreement in 1963. This involved a total of 25 million tons of crude oil to be delivered over five years, with the timing of supplies so arranged that by 1968 the Soviet Union would be supplying Italy with about 16 per cent of its total needs at price levels which gave discounts of some 30 per cent off posted prices. From Italy's point of view the bargain was even more favourable in that it also secured guaranteed export markets for massive quantities of Italian steel pipes and automobile factories, amongst other goods. Italy came under severe pressure from fellow members of NATO to modify the arrangement but it refused to do so and gradually increased its use of Soviet oil to over 12 million tons per year. To this degree of dependence on oil from the Soviet Union there was also added at a somewhat later stage the equivalent of another 5 million tons of oil per year in the form of Soviet natural gas. This will be discussed in more detail later in the chapter.

Only Britain and the Netherlands, amongst the principal Western European oil-consuming nations, continued for long to refuse to accept the advantages of Soviet oil – the latter because of its ownership of the larger part of the Royal Dutch/Shell Group and the former partly because of its interests in the ownership of two of the international major oil companies, but mainly because of its unwillingness to act against a NATO policy strongly supported by the United States. Britain finally reversed its policy in 1971 – when NATO ceased to argue against Soviet oil imports along its traditional lines, and some years after Shell itself had become much involved once again in the Eastern European oil trade. The Netherlands held out until 1974, when Soviet oil was imported to help break the Arab embargo on oil to Holland. Since then, however, much Soviet oil has flowed into Western Europe through the Dutch oil port of Rotterdam.

Soviet oil exports to Western Europe thus steadily increased from only 3 million tons in 1955 to over 40 million tons in 1969. The much increased oil prices on the world market since 1974 further stimulated Soviet oil exports to Western Europe so that, in spite of the fall in oil use in Western Europe since then and the rapid increase in indigenous production (especially from the North Sea), imports of Soviet crude oil and products have continued to increase. In 1984 the total was over 70 million tons: amounting to almost 20 per cent of Western Europe's total oil imports and to 14 per cent of total oil use. In response to this success and to ensure continuing success in the future, the Soviet Union has steadily improved its arrangements for delivering the oil to Western Europe (see Map 1, p. 54). At first the oil was moved from fields in the Soviet interior to the Black Sea ports by whatever means of transport were available, including pipeline where possible, and barges and rail as alternatives where necessary. In those early days the Soviet Union chartered tankers as required to ship the oil to Western Europe. Any 'profit' on its oil exports to Western European countries must have disappeared in such expensive transport arrangements and in others which were even more expensive. These included the rail-haul of fuel oils and even crude oil from the end of the pipeline system in European Russia to the Baltic ports of both the U.S.S.R. and Eastern European countries for transhipment to markets in Scandinavia. The demands of the increasing Western European export markets were, however, soon incorporated into the oil transport expansion programme under development by the U.S.S.R. and Comecon (the Soviet Union/Eastern European Organization for Economic Co-operation). These export demands did, in fact, help to justify the more rapid development of the pipeline system in the U.S.S.R. and Eastern Europe. Part of this now takes oil down to the Black Sea to ocean-going tankers whilst an even more significant part transports oil directly across European Russia into Poland and East Germany and so to the Baltic, where loading terminals for exports to Scandinavia and other western markets have been constructed. The demand for exports, over and above the demand for crude oil in Eastern Europe, justified the construction of this crude oil line which, with a diameter of over 100 cm, was then the world's biggest. A third line was constructed to take oil to Vienna (to supply Austria's main refinery) and to the Czechoslovakian Danube port of Bratislava, from where products could be transported by barge into southern Germany. Since the completion of these delivery facilities Soviet oil has had the advantage for Western Europe of being available through a more secure transport system than oil from any other part of the world. The completion of the pipelines also reduced the costs of supplying Russian oil to Western European markets and hence assisted its market penetration.

However, from the point of view of the Western European countries involved, Soviet oil remained only an alternative to importing oil from other sources. Thus purchases could be switched as contracts ran out in the light of changed economic conditions and of new political attitudes. Such uncertainty over export volumes was not, of course, very satisfactory from the Soviet Union's long-term planning point of view and it turned, therefore, to making efforts to secure outlets on a much longer-term basis. One effort was directed at persuading appropriately placed countries of Western Europe to allow extensions of the Soviet Union–East European pipeline system to be built to link with refineries in West Germany, Switzerland and even eastern France. With such direct deliveries, security of outlets would have been very much greater, but the efforts were unsuccessful, though in 1969 only the intervention of the Federal West German government, under severe pressure from the U.S.A., thwarted an agreement between the Soviet Union and the Bavarian state government. Had this agreement gone through, the Soviet Union would have been in a very strong position to put in branch pipelines to the other countries noted above. Later, in the context of the 1973 *rapprochement* between East and West Germany, there was a revival of interest in the possibilities of extending the Soviet–Comecon pipeline system to Western Europe. The first stage in such development was to be in conjunction with a refinery to be built in West Berlin and designed to run on Soviet crude oil, brought in by an extension of the line running to East German refineries. This plan, however, was also put on ice given the decline in oil use in West Germany as a consequence of much higher oil prices after 1974.

Along with the so far unsuccessful negotiations for extensions of its pipeline system to Western Europe, the Soviet Union also decided to try to secure firm outlets by buying or building refineries and distribution facilities in Western European countries. These efforts have been somewhat more successful. The Soviet oil export agency entered the market for petrol stations in the United Kingdom and so built up a distribution system fed by refined products exported to Britain from refineries on the Baltic and the Black Sea. Across the North Sea, in Belgium, the same agency successfully negotiated an interest in a refinery at Antwerp to which Soviet crude oil is fed by tankers from Soviet crude oil export ports and from which come products for distribution in Western Europe.

It thus became obvious by the early 1970s that the Soviet Union planned a long-term future for its oil exports to Western Europe. In order to secure this objective it first pursued commercial policies which were not dissimilar to those followed previously by American oil companies which had wished to break into these markets and so undercut prices when and where necessary to get the business which was sought. Somewhat later, however, after the 1973–4 oil supply crisis, the Soviet Union was able to take advantage of the great concern in Western

Europe over the security of oil supplies from O.P.E.C. countries. It then became possible for the U.S.S.R. to sell all the oil it had available for export to Western Europe without difficulties. It could do this, moreover, at prices related to the much higher price of international oil. By this time the earlier security-of-supply arguments against dependence on Soviet oil were emphatically reversed. This remained the case until recently when weak international oil market conditions arising out of Western Europe's declining use of oil have again tended to limit the rate of increase in the amount of Soviet oil sold in Western Europe.

Soviet oil exports to Western Europe have been made possible not only as a result of the increasing production of oil in the U.S.S.R., but also by the growing availability of natural gas as a primary source of energy for uses in the Soviet Union which would otherwise have required oil. Instead, the substitution of oil by gas in the U.S.S.R. itself has been deliberately planned in order to ensure an enhanced export potential of oil. By the late 1960s, however, it also became evident that the Soviet Union would still have a surplus of natural gas available and it began to seek export openings in Western Europe for this too. As with oil, only low additional transport costs are involved in getting this gas to Western Europe. This is because it can be moved through pipelines which were financed and constructed essentially to take gas from the producing areas of the Soviet Union to the consuming areas in the west of the country and to consumers in Eastern European countries. This was the main reason why Soviet gas – originating in fields over 5,000 kilometres away – had by the end of 1972 already achieved export markets in Austria, Italy, West Germany and even France, in competition with gas from the huge Groningen field of the Netherlands which, in this geographical context, virtually constitutes local production.

The first agreement was reached with Austria in 1968 for the delivery of upwards of about one milliard (10^9) cubic metres per year – the energy equivalent of almost 1 million tons of oil per annum. In 1969 first West Germany agreed to take 2.5×10^9 m^3 per year, and then Italy opted for 6.0×10^9 m^3 per year for a contract period of twenty years, making it the largest international gas contract ever signed up to that date. Negotiations with other Western European countries continued thereafter and were successful with Finland early in 1971. Agreement with France was reached in 1972 and at the same time West Germany and then Austria signed even larger contracts for additional Soviet gas to be delivered for twenty years from 1976.

These Soviet successes in exporting natural gas to Western Europe were initially a function of the policy pursued by the Dutch natural gas export concern – NAM – which aimed to maximize its short-term financial returns from gas exports, rather than the quantities of gas which it sold abroad. It thus fixed an export price which was far above the long-term supply price of the gas from the Groningen field and, in the absence of competition, stood to secure considerable monopoly profits. However, after 1971, partly as a result of competition from Soviet gas, Dutch pricing policy changed in order to meet the competition from the Soviet Union. There was also a NATO-sponsored idea that increased exports of Dutch gas should be offered at favourable prices in order to curb the flow of Soviet gas. Italy was specially favoured in this respect and a new pipeline was built to take Dutch gas through the Alps into the industrial area of northern Italy. These factors, for a time, made the Soviet Union's task of marketing gas in Western Europe somewhat more difficult than it had previously envisaged. Natural gas is some three times more expensive to transport by pipeline than the equivalent amount of energy in the form of oil and so the much nearer Dutch sources of gas had a significant advantage over Soviet supplies. However, much of the Soviet Union's natural gas was, at that time, a by-product of oil production, so that the costs of the producing operation could be set against the oil produced in the joint operation (as had traditionally been the case in the United States for much of its natural gas production); and this compensated for the disadvantages of the high transport costs involved in getting gas from the U.S.S.R. to Western European markets. As things turned out, however, this period of competitive conditions for Soviet gas exports proved to be very short. From 1973 – with the change in world oil markets – natural gas from all sources became a commodity with a very ready market indeed in Western Europe and the Soviet Union was able to increase both the volume it sold and the price at which it sold the gas.

Moreover, in 1973 the Soviet Union secured access to relatively low-cost gas from Iran, whose gas output from its main oilfields had hitherto been mainly flared at well-head. A large 100 cm line was constructed from the Iranian oilfields across the country to the border with the Soviet Union (see Map 1, p. 54). The line's capacity was over 15 milliard (10^9) m³ per year and it was originally made available at the border with the Soviet Union at a price of only about 2 U.S. cents per m³. With such a low purchase price and with only marginal transport costs to allocate to the gas moving to Western Europe (as fixed costs could be recovered from domestic and Eastern European sales), and with the prospects of even larger quantities of Iranian gas becoming available in the future, Western Europe seemed likely to find gas coming through the Soviet

65

Union from Iran an increasingly attractive proposition, particularly in the light of the greatly increased post-1973 price of oil and doubts about its security of supply. This prospect was, however, thwarted by the revolution in Iran, when the new régime suspended the supply of gas to the Soviet Union. As indicated above, however, it has recently been reinstated. It may even be expanded as Iran begins actively to seek export markets for its gas to help replace the revenue lost from reduced oil exports.

Meanwhile, the continued upward appraisal of the size of the gas resources of the Soviet Union itself (the U.S.S.R. is now thought to have over 36 per cent of the world's proven gas reserves), and the expansion of the natural gas market in Western Europe, led to an agreement in 1982–3 for the delivery of up to another 65 milliard cubic metres per year of Siberian gas to West Germany, Italy, Austria, France, Belgium and even the Netherlands. Although the United States has opposed this development – on grounds reminiscent of the arguments used in the 1950s and the 1960s against importing oil from the Soviet Union – the agreements seem likely, in the main, to be kept though the Netherlands cancelled its option to buy when its own reserves of gas were considerably upgraded. The new Soviet supplies are already being delivered in 1985; they will reach their plateau level in 1990 and will then flow at that level until at least the year 2005. This development will more than double the contribution of Soviet natural gas to Western Europe's energy supply and if, as expected, Soviet oil exports also continue at about their present level, then the U.S.S.R. is likely to be the single most important foreign supplier of energy to Western Europe by the end of the present decade.

Thus, given the geography of energy supply and demand in the region, even closer links in the energy sector seem likely to develop between the Soviet Union and Eastern Europe on the one hand, and Western Europe on the other. Though the development of such large long-distance overland energy flows may appear surprising, it should be noted that the distances involved are not much greater than those over which oil and gas are moved from the south-west of the United States to the important energy-consuming, industrialized parts of the country away to the north and north-east. And they certainly involve shorter distances and, seemingly, fewer problems than the planned movement of Alaskan and northern Canadian gas to the continental United States.

So from all angles, except the now largely discounted political one which favours minimum east–west contact, Soviet oil and gas links with Western Europe make good sense for both seller and buyer. Had the Soviet Union confined its oil-exporting activities to this area then suspicions about its motives might have been quickly dispelled. As things turned out, however, the Soviet Union either deliberately sought – or had thrust upon it – sales of oil in other parts of the world. Its success in selling to Japan – and the even greater potential which would follow the completion of a major pipeline all the way across Siberia to an export terminal on the Pacific coast of the Soviet Union (see Map 1, p. 54) – need not delay us very long. The motivations and the pattern of development, including the expansion of transport facilities, parallel what has happened between the Soviet Union and Western Europe and make as much economic sense from both parties' point of view – perhaps even more from the point of view of the Soviet Union, for its sales of oil to Japan enable it to buy, in return, the goods, capital and expertise needed for the economic development of Eastern Siberia and other Far East Soviet territories. The significance of this should not, moreover, be interpreted merely in economic terms, for the development and accompanying populating of these eastern territories could be as much connected with the need to strengthen the area's links with the rest of the Soviet Union, in the light of a possible future policy of expansion by China, as with economic growth pure and simple. Indirectly, then, the Soviet Union can be judged to be strengthening its defences *vis-à-vis* China partly on the basis of oil exports to Japan. This appears to give even greater justification for viewing Soviet interest in oil exports to Japan as essentially long term – a prospect which is made even more likely given the successes since 1977 of the joint Soviet/Japanese oil exploration activities in the offshore waters of Soviet Sakhalin. Should the finds which have already been made there herald the discovery of a major oil province, then the area's potential contribution to Japan's oil needs could eventually be significant. However, from the Japanese point of view this could raise political difficulties, given Japan's decision to seek some of the oil it requires from the expanding and potentially very large production in China – in a situation in which the state of Soviet–Chinese relations may not permit Japan to enjoy the benefits of importing oil from both countries (this point will be discussed at greater length in Chapter 6).

The question of the motivation for the U.S.S.R.'s oil-export policy arises mainly from the fact that over the last fifteen years Soviet oil has also been moving to parts of the Third World – to countries, that is, whose economic relationships with the Soviet Union can certainly not be interpreted in anything like the same way as those of Western Europe and Japan with the U.S.S.R. Most interpretations of the Soviet Union's willingness to send oil to countries like India, Liberia and Brazil (to give but one example from each of the three continents constituting the Third

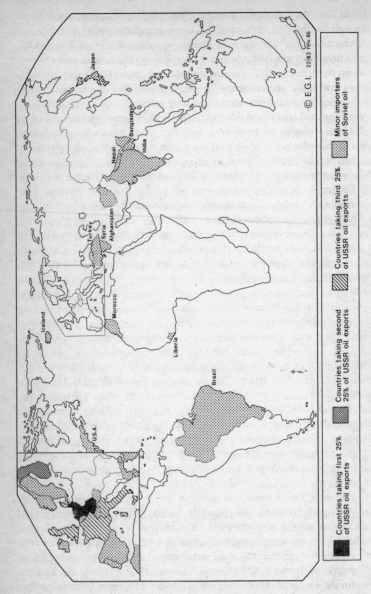

Map 2. Soviet oil exports to the non-communist world, 1984.

Countries taking first 25% of USSR oil exports

Countries taking second 25% of USSR oil exports

Countries taking third 25% of USSR oil exports

Minor importers of Soviet oil

© E.G.I. 22/83 rev 86

Japan

Bangladesh

Nepal

India

Turkey

Syria

Afghanistan

Iceland

Morocco

Liberia

Brazil

U.S.A.

World – see Map 2, p. 68) give pre-eminence to political motives. Seen through Soviet eyes such countries are in a position of economic sub-servience to the major capitalist powers, notably the United States, and it is a Soviet responsibility to help to break these bonds and, in doing so, realign the countries concerned with the socialist world. Offers of oil at prices lower than the countries can obtain within the framework of their neo-colonialist status provide an excellent means of attaining this aim at a very low real cost to the Soviet Union. In the short term, the relatively small quantities of oil required can be provided at little more than the cost of transport – a cost more likely than not to be easily recoverable in an acceptable currency from the country to which the oil is being sent. The validity of this argument from the Soviet point of view is un-challengeable. Indeed, as will be demonstrated in a later chapter, the developing countries did, in the main, find that they could get 'capitalist' oil only at prices far above competitive prices in the 1950s and the 1960s. This was, of course, the result of the control exercised over transport, refining and distribution of oil by a small group of international com-panies which, as shown in Chapter 1, had no incentive to break the 'unwritten code' governing the price at which oil was sold to such countries. To this degree Soviet oil sales to developing countries have been part of the conflict between the political systems of West and East. The most outstanding example of this arose in the case of Cuba, follow-ing Fidel Castro's successful take-over of power in that country in 1959.

Cuba produced practically none of its own energy requirements up to the time Castro took control. Its energy needs were instead imported in the form of crude oil and oil products from Venezuela by three of the major international oil companies. The new government of Castro began to question the validity of the import prices they were charged for these imports, given that they amounted on average to almost $3 per barrel, in spite of the low transport costs involved in getting the oil from Venezuela where at that time crude oil was available at the export terminals at under $2 per barrel. Castro's government, however, found itself unable to secure alternative lower-cost supplies in the absence of any importing and/or refining facilities outside the control of the major companies which claimed the absolute right to deny the use of their facilities to supplies from any source other than their own affiliated companies overseas. This legally enforceable right prevented Cuba from diversifying its sources of supply and thus from reducing the cost of its oil imports within the framework of the existing system. This adverse position for Cuba was confirmed when the companies refused to refine crude oil which the Soviet Union had agreed to sell to Cuba at a delivered price of $2·10 per barrel. Cuba's reaction was to interpret the refusal within the framework of its rapidly deteriorating relations with the U.S.A., whose government gave the oil companies responsible for Cuba's oil supplies a guarantee of its support for their refusal to handle Soviet oil. Indeed, the United States government may well have encouraged their action in order to bring its difficulties with Cuba to a head over a matter on which the 'law' was quite clear. The Cuban government refused to accept the companies' decisions and it countered by 'intervening' the refineries (that is, by placing them under direct government control). It then gave instructions for the importation and refining of Soviet oil to the extent of the trade agreement – enough to cover approximately one-third of the country's total requirements. The companies in their turn declined to have anything more to do with Cuba and withdrew their staff and terminated their supplies. The eastern political system then retaliated by agreeing to supply the whole of Cuba's oil needs – a decision that proved to be the catalyst which finally transformed Cuba into a member state of the international communist system. In terms of international oil trade it led to the establishment of the largest single flow of Soviet oil to any country outside Eastern Europe.

It would obviously be nonsense to deny the existence of basic political motivations in the Soviet Union's decision to sell oil to Cuba but only in the same way that it is impossible to exclude political considerations from the U.S.A.'s attitude towards oil sales there. Although the situation of rivalry over oil has not become so extreme elsewhere in the Third World, there are other developing countries where much the same elements have been involved. In these other cases more restrained reactions by the parties involved have inhibited such extremism. Sometimes the major oil companies concerned have agreed – albeit reluctantly – to handle Soviet oil through their refineries. In other cases governments have decided to provide their own oil handling and/or processing facilities, either by buying them from the companies concerned or by building new ones. And in yet other cases – most notably India, potentially one of the world's greatest oil markets – the 'threat' of Soviet oil as a replacement for oil imported by the international companies was sufficient to make the latter agree to significant price concessions.

It is, however, too facile to interpret this growing Soviet interest in the developing countries merely as part of the communist world's effort to subvert the capitalist world's system. Such an interpretation mistakenly equates the best interests of the developing countries necessarily and inevitably with continued trading relations in oil via the major international oil companies. In that these countries were, for many years, virtually held to ransom over such trade, because of their own weaknesses on both economic and political terms *vis-à-vis* the international oil companies and in the absence of effective competition between them, it is possible to argue that their economic interests have been better served as a result of the increasing availability of Soviet oil. The competition did at least give them some degree of choice and usually brought down the prices they had to pay for their oil from their traditional suppliers.

The wholly political interpretation also ignores the genuine Soviet economic interest in expanding its trade with developing countries. In spite of the great range of the Soviet Union's resources in foodstuffs, agricultural raw materials and minerals there are some of these commodities that it cannot produce itself in sufficiently large quantities – for example, copper, rubber and wool – the imports of which have, therefore, to be financed through exports. Oil is a good 'line' to offer for several reasons: first, because of the quantities in which it is available; second, because of the comparative advantage which the Soviet Union has in its trade; and third, because of its acceptability to developing nations as a substitute for supplies from other sources. Most other Soviet goods have specifications of unknown quality and in the case of machinery, motors and electrical equipment, etc., are generally not compatible with the existing stocks which, in the case of most countries of the Third World, originate from North America, Western Europe, and/or Japan.

The Soviet Union has yet a further economic incentive to increase its trade with developing countries. This arises from the ability of many of these countries to supply tropical and sub-tropical products such as cocoa, rice, coffee and fruits, the domestic production of which the Soviet Union's physical environment prevents or severely limits. Until recently such commodities were considered luxuries which neither the Soviet Union as a country nor most of its citizens as individuals could afford. Increasing affluence, however, has created a vast potential demand for such products, some part of which can be effectively met as a result of bilateral trading developments arising out of oil exports. Thus there are some strong economic factors at work encouraging the export of Soviet oil to the developing countries and the trade can be expected to continue to grow insofar as the Soviet Union has oil to spare for such trade. This is by no means self-evident given the limits on Soviet oil production and the fact that the U.S.S.R. will continue to give preference to oil exports to countries of both Eastern and Western Europe for sound economic reasons. However, given the fact that so many developing countries have, over the last ten to twenty years, secured a higher degree of independence in their oil sector, by obtaining control over their own refining and distribution facilities (a topic which is dealt with at greater length in Chapter 7), the prospects for markets for Soviet oil exports are thus enhanced. In this respect it is worth noting specifically that the Soviet Union and its allies in Eastern Europe have helped to create this degree of independence through the aid which they have given to the oil industries of the developing countries. Soviet statements have indicated time and time again the importance the U.S.S.R. attaches to such aid. When one notes that almost 25 per cent of the finance for the oil sector in India's third five-year development plan was provided by Soviet or allied sources, or that the Soviet Union agreed to provide experts and finance for the exploitation of Brazil's oil shale resources, or that more than a dozen other developing countries have accepted Soviet assistance for oil developments of one kind or another, then it is clear that the communist world's ability in this direction should not be understated.

Some years ago, W. E. Pratt, a leading petroleum geologist and engineer, and formerly Vice-President of the Standard Oil Company of New Jersey, claimed that 'American freedom of enterprise appears to be indispensable to the task of finding oil in the earth'. He went on to say that 'Russia has vast territories ideal in their promise for new oilfields', and that 'situated in the United States and explored by those American methods, the same territories would pour out a veritable flood of oil'. Within the space of just a few years the Soviet Union has achieved that 'veritable flood' without the 'benefits' of American men, methods or capital, and its oil industry has reached levels of production which were considered highly unlikely, if not impossible, by American experts only a few years ago.

It could be construed that it was the somewhat reluctant recognition by the United States of the achievements of the Soviet oil industry and the potential dangers arising from that success which led to the proposal for cooperation in the field of oil and gas between the U.S.A. and the U.S.S.R. as one element in the political détente achieved by the two super-powers in the early 1970s. Agreements in principle for the use of American capital and technology in the rapid exploitation of known, but as yet undeveloped, fields in Siberia and for Soviet oil and liquefied natural gas exports to the United States were signed in June 1973. The agreements, however, were not implemented, partly as a result of the time which was required for working out the details, and partly because of their necessary re-evaluation by both parties in the light of the fundamental changes in the world oil power position after the events of October 1973. For example, the agreement that the U.S.S.R. should supply energy to the United States had to be looked at again in the light of the latter's post-1973 aim to re-achieve energy self-sufficiency. Trade in liquefied natural gas, if not in oil as well, from the Soviet Union to the United States nevertheless still seemed a possibility for later in the 1980s in spite of the semi-freeze on developing American–Soviet economic contacts in the aftermath of the Soviet invasion of Afghanistan and the problems over Poland. It is noteworthy, in this context, that the 1981 ban on the export of American technology to the U.S.S.R. originally excluded oil and gas industry equipment and expertise. Meanwhile, however, the declining use of oil in the U.S. has undermined American interest in the exploitation of Soviet resources. More recently American policy under President Reagan reverted to one which excluded cooperation with the U.S.S.R. so that there was even U.S. opposition to Soviet gas exports to Western Europe and to Western European countries' sale of oil industry hardware to the U.S.S.R., let alone to direct U.S. imports of Soviet oil and gas. These policies of the Reagan administration constitute a splendid example of the way in which decisions on oil and gas are inextricably interwoven with attitudes on general political and economic issues.

Over the rest of the century it now seems well nigh certain that the Soviet Union will move even further ahead of the U.S.A. in oil and gas production, in order to satisfy its national and its allies' demands, and to develop new markets in other parts of the world. In other words, United States' Central Intelligence Agency reports in the latter 1970s/early 1980s on an impending Soviet shortage of energy seem unlikely to be correct. The conclusions of those reports seemed more designed to 'frighten' U.S. consumers into more oil conservationist behaviour than to give an objective analysis of Soviet opportunities for the continued development of its resources. If during this same period the U.S.S.R. also decides to involve itself to an even greater extent in oil and gas production in the Middle East – as now seems likely, given its agreements with Iraq and Libya and its more recent greatly enhanced general interest in the region – then such involvement will emerge in part out of growing Soviet expertise in this field, of both a technical and commercial character, as well as from political interference for its own sake, or out of an overall Soviet need to supplement its own energy resources because of the lack of developments of – or the inability to go on developing – its own potential. The Soviet Union has demonstrated that an oil industry working on other than capitalist lines can be successful, and it will wish to continue to demonstrate this in other parts of the world, including the Middle East, in much the same way that American companies, proud of their heritage, have exported their way of doing things. Commercially, the Soviet Union will thus also gain access to oil on which it can make a 'profit' by selling it – or a substitute for it (e.g. more gas instead of oil) – in the world's markets. Here again, its behaviour would be in line with good American oil company practice. In other words in one way or another the influence of the Soviet Union based on the expansion of its oil and gas industries will continue to grow through the rest of the century. This expansion will, on the one hand, continue to create opportunities for trade but, on the other, it will also continue to cause consternation in the western world over the possible implications of such an expansion of Soviet oil interests in an increasing number of countries hitherto considered the preserve of the mainly U.S.-backed international oil companies.

4. The Major Oil-Exporting Countries

The main area of activity of the international oil industry lies outside the U.S.A., the U.S.S.R. and the world's other communist countries. This area comprises two sets of nation states. The larger set consists of the hundred and more nations in the world's increasingly complex political framework whose interests in oil and whose attractions for the international oil industry lie in their functions as oil-consuming nations. These countries' oil problems, and their associated geo-political implications, will be considered in Chapters 5, 6 and 7. Meanwhile, in this chapter our concern is with the much smaller set of nation states. This comprises a handful of countries which produce oil not primarily for their own use, but rather for export, in crude of refined form, to other countries of the world. These nations are picked out in Map 3 (p. 76), which shows those countries in which oil production exceeds oil consumption by four times or more.

Not only are there few of these – only twenty in all – but as the map shows they are geographically concentrated in the Middle East and Africa. For both these reasons, it is highly likely that, whenever oil is in the press headlines, it will be the result of a political or economic change in one or other of these countries or because of a geo-political problem arising out of the relationships of these countries with each other or with the rest of the world. The interests of the oil-producing countries are clear and easily recognized, especially as, in almost every case, the oil industry forms the dominant element in their economies. Thus, negotiations between these oil-exporting countries that take place from time to time are concerned with the fundamental national interests of the countries involved, and raise great issues which are aired at considerable length in the media. This phenomenon has never been more clearly marked than in the period since October 1973, when the oil-exporting countries began at last effectively to exercise their potential influence on world politics and economics for the first time. Without oil these countries would certainly be of significantly less importance on the world stage, and far less attention would be paid to the Middle East – the area in which many of the oil-exporting countries are concentrated. However,

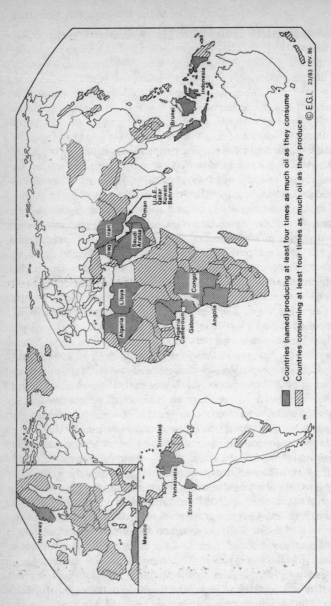

Map 3. The world's oil-exporting and importing countries, 1985.

Countries (named) producing at least four times as much oil as they consume

Countries consuming at least four times as much oil as they produce

U.A.E.
Qatar
Kuwait
Bahrein
Oman

Iran
Iraq
Saudi Arabia

Indonesia
Brunei

Algeria
Libya

Nigeria
Cameroun
Gabon
Congo
Angola

Norway

Mexico
Trinidad
Venezuela
Ecuador

© E.G.I. 23/83 rev. 86

the relatively recent rise of the Middle East to importance in this respect should be noted. It is essentially a post-Second World War phenomenon. Before the war Middle East countries were involved to only a very limited degree in the supply of oil to the rest of the world. The latter, in any case, was then much less interested in oil owing to the greater importance of other fuels, and the lower all-round consumption of energy as a result of the lower degree of economic development and the relative unimportance of road and air transport. Before the war the world's single most important oil-exporting nation was the U.S.A., with the U.S.S.R. and Mexico not far behind. Today, their functions as overseas suppliers of oil are limited in relation both to their own use of oil and to total world trade in the commodity. Instead, Venezuela, various countries in the Middle East – especially around the Persian Gulf – several North and West African states, and a few other countries elsewhere have rapidly increased their contribution to total world oil supplies. From 1945 to 1974 they also provided, year by year, an increasing share of the world's total energy requirements. Even though their relative importance has since declined, the nations we are concerned with in this chapter still account for about 60 per cent of total world oil production outside the U.S.A. and the U.S.S.R., and for more than 75 per cent of all oil moving in international trade.

The first nation to undergo a meteoric rise to significance as a major producer and exporter was Venezuela. This occurred in the 1940s after twenty years of somewhat desultory oil exploration there by a number of the international oil companies. They were finally galvanized into an urgent flurry of activity as a result of the expropriation of their assets by, and their expulsion from, Mexico, where the oil industry was brought under national ownership in 1938. For twenty-eight years a succession of revolutionary governments in Mexico had always seen such action as the ultimate outcome of the conflict between state and oil companies but, since they had avoided nationalization for so long, the companies had come to believe it would never happen and thus they paid relatively little attention to developing large-scale alternative production potential elsewhere. When they finally lost their Mexican oil, the promising prospects for the exploitation of oil in the Maracaibo Basin and in other parts of Venezuela benefited from the needs of Shell and Esso – the main companies involved in Mexico – to develop enough oil-producing capacity as quickly as possible to replace the 15 million tons or so per annum they had been lifting from their Mexican fields, mainly for sale overseas. This important stimulus to Venezuelan oil development was, moreover, soon supplemented by a second and equally important one,

77

viz. the rapidly expanding petroleum needs of a wartime U.S. economy. These wartime demands made it impossible for the U.S. domestic oil industry to meet the oil needs of the United States itself, of allied countries and of the armed forces around the world. This situation gave an increased number of companies a still greater incentive to seek new resources in Venezuela. Venezuela was the nearest possible external source of oil supply for the United States (except for Mexico, whose oil was embargoed by the U.S. as a result of its nationalization of the oil companies), and so its supplies were at a minimum possible risk from enemy action at sea. As a result, oil production in Venezuela rose rapidly from only 20 million tons in 1937 to some 30 million tons in 1941, and to over 90 million tons by 1946, by which time the country was the world's most important petroleum-producing nation outside the United States. Since almost all the oil was exported, in contrast with the mainly domestic use of American oil, Venezuela became the world's most important oil exporter (a position which it held on to for over two decades until 1970, but which position it then lost to Iran and Saudi Arabia).

In the post-war world, in which there were fairly general difficulties over energy supplies as a result of dislocation in many of the most important coal-producing areas, the demand for energy from alternative supply points such as Venezuela grew rapidly. The Venezuelan politico-economic environment at the same time was also highly favourable to foreign investment in oil because the dictatorial régime there welcomed such investment as a means whereby those individuals close to the régime could amass private fortunes. These two factors ensured the continuation of the growth of Venezuelan oil production and exports throughout the rest of the 1940s and up to 1957.

The twenty-year period of growth in the oil industry in Venezuela was marked by only one short interlude of restraint – during the few months in 1948 when a government came to power under the leadership of a political party, Acción Democrática, whose electoral manifesto called for the nationalization of the country's oil resources and whose leaders in exile had lived mainly in Mexico, where the oil industry, as indicated above, had been nationalized in 1938. The reaction of the oil companies to this new Venezuelan government was immediate and very blatant. Investment virtually ceased, development came to a halt and production was stabilized while the top management of the international companies concerned attempted to decide how far they were prepared and able to work within the framework of the policies likely to be adopted by the new régime. As it turned out their fears for their investments and the supplies of oil from Venezuela for their international markets were shortlived for, after a short period of democratic rule, the country reverted to a military dictatorship – a reversion to the *status quo ante* which was almost certainly only made possible so quickly with the active help of at least some of the oil companies involved in Venezuela.

The country's new dictatorship of Pérez Jiménez soon regained the confidence of the oil companies, and oil production was once more established on its strong upward trend. Companies like Shell and Esso, long established in Venezuela, and also other companies new to the country, but anxious to participate in the bonanza, fell over each other in their efforts to secure concessions in the large, promising areas of Venezuela which were still unexplored. In the mid-1950s Pérez Jiménez organized what amounted to a 'grand auction' of some of these areas. The successful bidders paid some $1,500 million (equal to over $6,000 million in 1985 values) for the exploration and development rights and started work immediately to prove that Venezuelan production could be pushed beyond the annual level of over 150 million tons already achieved. No sooner had this work got under way, however, than the dictatorship of Pérez Jiménez in Venezuela was overthrown and, after a short period of provisional rule, Acción Democrática again achieved power in 1958. Predictably, the reaction of the oil companies developed at first in much the same way as it had on the earlier occasion of an Acción Democrática government in 1948. Once again investments and development plans were curtailed or abandoned and oil offtake limited, as the companies sought clarification of the possible oil programme of the new government and evaluated the strategies open to them. This time, however, the rule of the Acción Democrática government was not shortlived. In fact, it achieved what no other elected government in Venezuela had ever achieved: it served its full term of office and it was succeeded by another elected government.

Nevertheless, in 1958, at the beginning of its term of office conflict between the government and the companies seemed almost inevitable, as Acción Democrática still had proposals for the nationalization of the oil industry in its manifesto and, moreover, it took action early in its period of government steeply to increase taxes on the industry. It also gave its support to the oil industry trade union's pressure for greatly increased wages and fringe benefits. These developments seemed to indicate that a head-on clash was but a matter of time. But after its election Acción Democrática did not treat its nationalization commitment seriously and certainly made no move in that direction. In fact by 1958 Venezuela was so completely dependent economically on the oil industry that no government – and certainly not one as anxious as Acción Democrática to enhance the country's economic progress – could afford to think of action which would essentially close down the oil sector of the economy for at least a number of years. No other sector would have been unaffected by the repercussions of such action, and the consequent unemployment and distress would certainly have undermined the government's political strength. The government's freedom of action in economic terms was thus heavily constrained and, even in political terms, there was little to be said for action which, no matter how immediately 'popular', seemed likely to create such stresses and strains in the system that the instigators of it were unlikely to survive.

But if by 1958 the government's ability to act out its basic philosophical beliefs in respect of the organization of the country's oil industry was constrained, then so was that of the oil companies. By this time, as explained previously (see pp. 29–30), they were under pressure from the U.S. State Department to achieve a *rapprochement* with Venezuela, which was believed by the United States to be the country which provided the key to the stability of the whole Caribbean area. Stability in Venezuela, however, – particularly in the period following Fidel Castro's success in Cuba – demanded an expanding economy. This, in turn, depended upon the continuing development of the country's oil industry which, in the late 1950s, accounted for something like 25 per cent of the country's gross national product, provided the government with over 60 per cent of all its revenues and accounted for over 90 per cent of the nation's total exports. The companies, therefore, though powerful in the Venezuelan context, had to reorientate their attitudes and policies to the even more powerful force of the foreign policy needs of the U.S.A. These required that the oil companies make it possible for Venezuela to achieve its objectives of continued economic advance. This demanded the companies' willing cooperation with a government which they certainly disliked and probably distrusted, but for which there was no acceptable alternative and which therefore they could certainly not think of overthrowing as they had done in 1948. Economic and political necessities, therefore, as set out by the U.S.A., produced a situation in which the international oil companies dedicated to the idea of as little government intervention in industry as possible, and a government, devoted in theory at least to socialist planning, had to work together. This development – which was unusual for its time – has since been paralleled in both oil-producing and oil-consuming nations as the companies have been obliged to recognize the validity and permanence of governmental concern over oil and oil policies. Some other cases of this kind are, indeed, presented in both this and later chapters.

Expansion in Venezuelan oil production after 1958 was by no means as rapid as in the earlier post-war period. Nevertheless, advances did take place as investment, though on a smaller scale, continued. Government revenues from oil thus continued to increase. This occurred, moreover, in spite of the fact that in the period between 1958 and 1973 Venezuelan oil became increasingly uncompetitive in many markets of the world as a result of the rapidly expanding, lower-cost oil output from countries in the Middle East and, somewhat later, from North and West Africa. Moreover, falling unit costs of transporting oil across the oceans – as larger and larger tankers were brought into use – helped to eliminate Venezuela's competitive edge in the markets which were closer geographically to it than to other main producing areas. This was particularly important with respect to the U.S. market, which had hitherto been considered the particular preserve of Venezuelan oil, but in which Middle Eastern and other oil now became increasingly competitive.

However, it is necessary to stress that the slower rate of growth of Venezuela's oil industry was not only a result of its deteriorating competitive position in overseas markets. The new government of the country did not accept the need for, or the desirability of, as rapid an expansion of the oil industry as had its predecessors. Two arguments were advanced in support of this new policy. First, Acción Democrática considered that Venezuela had become too dependent on oil and that this was a threat to the well-being of the country in the long term. It therefore deliberately took action to reduce the oil industry's share of the total economy by encouraging diversification into other fields. Secondly, the government argued that rapid expansion of oil production and sales was at the expense of the price obtained for each barrel of oil. By restricting output, the government argued, price levels could be maintained and Venezuela's oil could thus be conserved to provide the means of economic development in the future. This early 'conservationist' view on oil-development strategy depended for its validity on several assumptions: first, that there existed a very limited amount of oil for the world's future use; second, that oil would not be replaced to any large degree by other forms of energy in the future for which the oil was being 'conserved'; and third, that only inadequate use could be made of the wealth created by additional oil sales in the short term. None of these assumptions was, however, valid at the time and thus the government's argument that Venezuelan oil should be conserved was weak. In addition, however, the argument was undermined by the unwillingness at that time on the part of other oil-exporting countries in competition for markets with Venezuela to accept that restraint on production levels was a feasible policy option. Thus, Venezuela's idea of protecting the price level of its oil through the pursuit of a conservationist policy was doomed to failure. Other exporting countries would simply have moved in to supply the oil that Venezuela might otherwise have had the opportunity of supplying. Thus, the deterioration in prices would continue irrespective of Venezuela's unwillingness to supply additional oil so that the net result for Venezuela would have been the loss of part of its income. Venezuela's enthusiasm for the creation of an organization which would enable the producing countries jointly to formulate and implement policies to defend their common interests stemmed from this situation. The formation and work of the Organization of Petroleum Exporting Countries (O.P.E.C.) was largely the result of Venezuela's efforts in this direction. The development will be discussed later in this chapter.

From the interplay of all these economic and political forces, Venezuela achieved an average annual growth rate in oil production of rather less than 3 per cent between 1958 and 1973. This compared with the 10 per cent per annum growth rate which Venezuela had achieved over the previous fifteen years, and a world annual growth rate in the fifteen years from 1958 which averaged more than 7 per cent. In 1958 Venezuela's production of 142 million tons accounted for 17 per cent of world total; its 1972 production of 182 million tons was only 7 per cent of the total and this was in spite of the fact that the closure of the Suez Canal since mid-1967 had given Venezuelan oil a boost in markets west of Suez, particularly in the United States. Even this locational advantage for Venezuela had, however, been undermined by the rapidly falling costs of shipping Middle East oil to the U.S.A. around southern Africa, instead of through the Suez Canal. Thereafter, it was not until the 1973–4 Arab embargo on oil supplies to the United States that Venezuela was able to reassert its ability to charge relatively higher prices for its oil exports. This persisted through to 1978, though with decreasing justification, as shipping rates fell still further and as the improvement of its ports and the construction of new facilities for the transhipment of crude oil in the Caribbean enabled the U.S.A. to take advantage of the transport cost savings which could be secured by using larger tankers for bringing in African and Middle East oil. As a result Venezuela seemed likely to have to reduce its prices relative to supplies from elsewhere if it was to achieve its objective of maintaining exports at about the 100 million tons a year level. Events in Iran, however, which severely reduced that country's exports in 1979, enabled Venezuela to keep up its export prices for a little longer, but the development of a situation of too much international oil chasing too few markets with the sharp fall in oil use after 1980–81, eventually meant that Venezuela had to adjust its prices downwards in order to achieve the level of exports which its economy needed to survive. By this time, moreover, the 'natural advantage' of Venezuela as the sole oil exporter of importance in the western hemisphere was being undermined by the rapid enhancement of Mexico's potential as an exporter. This specific competition further exacerbated Venezuela's difficulties in the international market.

Prior to further consideration of the most recent developments in the position of the Venezuelan oil industry *vis-à-vis* that of the rest of the world, it is necessary to look rather more closely at the evolution of the industry in the 1960s and 1970s. Though the Cuban crisis and resultant pressures by the U.S. State Department can be seen as the main factors which saved the Venezuelan oil industry from serious decline in the 1960s, one must also note the impact of the growing professionalism of the Venezuelan government in dealing with the oil companies. In earlier days the expertise was all on the side of the latter which, therefore, had to respond only to political pressure from the government rather than argue the technical and economic validity of their decisions. After 1958, however, the Ministry of Mines and Hydrocarbons in Venezuela built up a team able to argue in technical and economic terms with the industry's representatives and, as a result, it was often able to offer advice to the government as to exactly how much pressure should be put on the companies to make concessions – particularly in respect of taxation arrangements. Thus, the government was able to increase its share of total oil industry profits on several occasions; to collect taxes in arrears, the liability for which the companies challenged; and to change the basis on which total profits were calculated by obliging the companies to write into their calculations a notional tax-reference selling price which was, until 1973, invariably higher than the actual selling price of Venezuelan oil. All this had the effect of increasing the revenues which the country collected on every barrel of oil exported, and by 1970 revenues amounted to more than $1 per barrel, compared with less than half this amount when Acción Democrática came to power in 1958. This was achieved, moreover, despite the generally weak market for Venezuelan oil throughout the period and which led to export prices declining even in current dollars (and even more so in real terms). By virtue of its actions the Acción Democrática government succeeded in ensuring that its revenues from oil continued to grow at a rate high enough to finance the requirements of its ambitious economic and social development programme – the main short-term aim of the government in respect of its policy towards the oil sector.

Limitations to the continued success of this approach were becoming very obvious by the end of the 1960s. Government revenues per barrel could not be pushed to a higher level without eliminating all the profits to the companies concerned and if this happened then the companies would no longer have any incentive to continue to produce oil in Venezuela, unless the prices at which Venezuelan oil was sold overseas rose. In the absence at that time of prospects for oil price increases, increases in government revenues depended upon expanding the output of oil. But this was something that could not be achieved through the existing concession system, even if the government had wanted it to happen, because the oil companies had become unwilling to invest the necessary money in Venezuela, given more rewarding opportunities for their funds elsewhere in the world of oil. In any case, the government did not accept the idea of the concession system as a means of producing the nation's natural resources, except as a short-term expedient to ensure the continued flow of oil and, in the light of external pressures particularly from the United States, as discussed above, in order to allow the existing concessionaires to work their agreed areas. Since 1958, therefore, no new concessions had been granted, and as a result Venezuela's proven oil reserves had declined to little more than about thirteen years' supply at the annual rate of production in the late 1960s. Thus, in the absence of any change in the situation Venezuelan oil output would soon inevitably start to decline, so that, by the time the oil companies had legally to relinquish their concessions in 1983, it seemed likely that Venezuela could be little more than a relatively minor producer of oil.

In line with its political philosophy, Acción Democrática first sought to solve this potential problem through the establishment of a state oil company – Compañia Venezuelano del Petroleo (C.V.P.). This was given the responsibility for working any concession areas which might be relinquished by private companies and for negotiating joint arrangements with foreign oil companies, etc., to work as yet unexplored areas of Venezuela with oil potential. It made a start in a small way as far as the former responsibility was concerned and achieved a producing capacity amounting to about 9 million tons per year. And in respect of the latter responsibility, it accepted offers from a dozen or so petroleum companies for joint operations in the southern part of Lake Maracaibo. The work in this area was to be carried out and financed by the companies, but the ownership of any oil discovered would be vested in the state entity, which would then have to pay a per barrel fee to the successful companies for their efforts. The companies, in turn, would have the right to purchase so much of the oil – if they so wished.

This contractual system between the state oil company and private foreign companies did, in theory, give the Venezuelans a much greater direct control over the development of their oil resources than the concession system it replaced. Some foreign companies felt justified in seeking to participate in the new system, so that given time it could well have prevented an early and possibly catastrophic decline in production levels, thus safeguarding government revenues for a few more years. However, it was quickly overtaken by events and then abandoned in 1976 in favour of a quite fundamentally different approach to the ownership and development of the oil industry in Venezuela.

By 1976, of course, the major oil-exporting countries including Venezuela had, through O.P.E.C., taken control over the price of oil and, as a result, were securing government revenues of about $10 on each barrel of oil exported. Even after allowing for inflation since 1973 this represented a more than fivefold increase in the value of such revenues and this placed Venezuela in a very much stronger position to treat with the oil companies. Formal discussions in Venezuelan political and economic circles on the implications of the changed relationships with the companies began in 1974 with the appointment of a national commission to review the position, role and organization of the country's oil industry. This Commission recommended the nationalization of the industry, basically on the grounds that the national interest necessitated ownership and control over oil as the commanding height of the Venezuelan economy. It also recommended that nationalization be carried out immediately in order to avoid the problems for the country which were being created by the unwillingness of the foreign oil companies to invest any more funds in Venezuelan oil developments because their concessions were due to be terminated in 1983-4. In these circumstances the companies could not reasonably expect to achieve any financial benefits from such new investment. They were interested for the last few years of their concession agreements only in the returns they could make from the capital they had invested in previous years. From the country's point of view, however, this was a totally unacceptable situation and hence the recommendation by the Commission for an immediate state take-over of the oil industry was accepted. It was implemented with effect from 1 January 1976.

The oil companies were compensated, but at a level related only to the rundown 'book-value' of their assets which were determined, though not with the agreement of the oil companies, at no more than $1,000 million, given the absence of much by way of new investment for more than a decade. Even this level of compensation was, moreover, offset in part by

85

the costs incurred by Petróleos de Venezuela – the new state oil company – in bringing the industry's physical assets back into full working order. This was done, again without the companies' agreement, on the grounds that the companies had failed to meet the terms of their concession rights in allowing their assets to fall into disrepair. Thus, Venezuela achieved the national ownership of its oil industry at a modest cost (though how modest remained a matter of dispute for some years as the companies took the question of the level of compensation to the courts) and in doing so brought the era of 'oil imperialism' to an end – at least as far as the situation internal to Venezuela was concerned.

However, in terms of its international relationships – in respect, in particular, of the sale of Venezuelan oil abroad – Petróleos de Venezuela remained, of course, heavily dependent on the arrangements it had to make with oil companies, such as Exxon, Shell and Gulf, which formerly operated concessions in Venezuela. This was because these companies 'sold' most of their Venezuelan production to their sister companies in other countries for refining and distribution and these outlets remained necessary as markets for most Venezuelan state oil. In 1977 the state company only managed to find independent markets for about 20 per cent of Venezuela's exports, so the success of the national oil industry depended in the early post-nationalization years on its agreements with the former owning and operating companies. The ability of Petróleos de Venezuela to sell oil abroad gradually increased as it succeeded in building up an international marketing organization. Nevertheless, it has been a fairly slow process, particularly as competition from the international companies remained strong, and the state entity did not wish to offer oil at heavily discounted prices for fear of undermining O.P.E.C. agreed price levels, and so possibly starting a price-cutting war in which government revenues would decline to the disadvantage of Venezuela. Petróleos de Venezuela avoided this danger in the three years after its formation – and then in 1979 the danger disappeared as oil became short on the international market consequent upon the massive decline in Iranian exports. Thereafter state oil sales could, for a time, be made at a premium – even to the companies which formerly ran the Venezuelan oil industry. Since 1981, however, the over-supply of oil on world markets, particularly heavy oil of the kind which constitutes most of Venezuela's exports, has undermined the sales efforts of Petróleos de Venezuela which has thus again had to seek cooperation with the international oil companies. The latter, now in a stronger position, linked their willingness to help to the achievement of a solution to their claims over compensation for their nationalized assets and to the government's counter-claims for the companies' earlier alleged under-payment of taxes, showing clearly how international oil power considerations can influence national developments.

What has happened in respect of the ownership and organization of the oil industry in recent years in Venezuela is largely a consequence of the post-1973 changes in the world of oil power; and it closely parallels what has happened in almost every other major oil-exporting country. We shall return to describe and to evaluate the fundamental changes in oil power since 1973 later in the book (Chapter 9). In the meantime we must turn our attention to the interest and policies of the world's even more important oil-producing and exporting region – the Middle East.

Middle Eastern oil production dates back almost seventy-five years. As early as 1913 the importance of oil from Persia for Britain, in particular, was demonstrated in the decision of the British government to finance the continuation of exploration there at a time when private interests were considering giving up their efforts. This action created the Anglo-Persian Oil Company, in which the British government held a 51 per cent interest, an interest which was maintained until 1979, except for a short period in the early 1970s when it fell to about 48 per cent. Over the years the designation of the company has changed twice, first to Anglo-Iranian in 1935 and then, in the early 1950s, to British Petroleum. As a result of its close ties with Britain (the most influential western power in the region from the mid nineteenth century until after the Second World War), Persia (Iran) became the major oil-producing nation in the Middle East. It maintained this position throughout the inter-war period and thereafter until it nationalized its oil industry in 1951. In 1939 it was already producing about 10 million tons per year, twice as much as was then being produced in all the other Middle Eastern countries put together. Meanwhile, disagreements between Britain, France and the U.S.A. over the political control and the development of the oil resources of other parts of the Middle East had delayed the expansion of production in other parts of the region – as, for example, in countries such as Kuwait and Iraq (see Chapter 8 for a discussion of this situation) – but when an agreement between the external powers on their respective areas of influence was finally signed in 1935 the scene was set for the exploitation of the Middle Eastern countries' already known oil resources. In the few years remaining until the outbreak of the Second World War development went ahead quickly in Iraq, so that the country's output increased to about 4 million tons in 1939. The Anglo-Iranian Company and the U.S. company, Gulf Oil, which had jointly, on a fifty-fifty basis, secured the concession to explore for and to exploit the oil resources of the sheikdom of Kuwait, soon discovered an oilfield which promised to be of major significance. Unfortunately, the outbreak of war in 1939 made the development of the field impossible. As a result, Kuwait's entry into the ranks of the oil-producing and exporting nations was delayed for almost a decade.

During the Second World War production was maintained as far as possible from those fields in the Middle East that had already been developed by 1939. Their production was needed in order to meet the demands of the Allied forces east of Suez, as well as those of civilian markets in areas around the Indian Ocean and in the southernmost parts of the U.S.S.R. to which oil was exported from Persia (Iran). But such a limited demand acted as a restraint on the growth of the area's oil industry during this period, and in 1944 – the last full year of the war in Europe and the Middle East – the whole of the Middle East produced only 20 million tons. This was little more than oil production in 1939, and by the end of the Second World War the Middle East had fallen way behind Venezuela as a major producing region.

The immediate post-war period brought rapid changes. The seven international companies quickly started exploration and development work throughout the Middle East under the same stimulus that had encouraged them (with the exception of B.P.) to pursue their active efforts in Venezuela; that is, a world short of energy and hence a world willing to pay high prices for any oil that could be made available to the markets of Western Europe and elsewhere. Moreover, oil produced from the Middle East was particularly profitable as it could be sold in the early post-war years at a delivered price which not only reflected the higher costs of producing oil in the U.S.A. but which, for most customers, also included a transport cost component calculated as if the oil originated in the Gulf of Mexico. For most markets of the world to which Middle East oil was delivered this meant that higher freight charges could be levied than were necessary to cover the costs of actually transporting it from the Middle East. This commercial advantage arising from supplying Middle East oil to secure markets was further enhanced by the companies' knowledge that the area's political stability was still guaranteed by the presence of Britain in both political and military guises throughout the region. Thus in the early post-war years the companies had no hesitation in committing many hundreds of millions of dollars to the exploitation of the Middle East's prolific oil resources.

The two main oil-producing countries of the region, Iran and Iraq, quickly benefited as output from their existing oilfields was increased by investment in the producing operations themselves, and by expenditure on increasing the capacity of the pipelines and other facilities needed to get the oil from the fields to markets (see Map 4, pp. 90–91). Further south, the pre-war discoveries in Kuwait were speedily developed and even the earlier very optimistic evaluation of its fields was quickly shown to be much too low. In Kuwait, it seemed, the companies could hardly put a drill in the wrong place, for almost every 'wildcat' and development well that was sunk proved to be a producer of an enormous size, with many wells capable of producing up to 10,000 barrels per day, the energy equivalent of about three-quarters of a million tons of coal per year. The two companies with shares in this bonanza – Anglo-Iranian and Gulf Oil – very soon became embarrassed by the size of the resources with which they had to cope because neither of them had large enough markets for such quantities of oil within their own marketing companies (unless Kuwaiti oil was used merely to replace production from elsewhere – Iran in the case of Anglo-Iranian and the U.S.A. in the case of Gulf). Both companies, therefore, made long-term arrangements to sell large quantities of Kuwaiti oil to other major companies with marketing opportunities in markets which were growing rapidly and so able to absorb Kuwait's production. Under the stimulus of these bulk sales to Shell and to some of the large American companies, Gulf and Anglo-Iranian were able to continue to expand their output in Kuwait.

Meanwhile, in Saudi Arabia, from which British oil interests had been excluded as a result of U.S. diplomatic pressure and activities, a consortium of American companies only was formed, under the name of Aramco, with concession rights to discover and exploit the oil resources of practically all that vast country. Again success was immediate and cheap, and the fields of Saudi Arabia were brought into production as quickly as the physical facilities needed to move the oil to the coast for tanker transportation to the markets of Western Europe – and, in increasing quantities by the late 1940s, to the U.S.A. – could be completed.

From these early post-war developments, production in these four Middle Eastern countries continued to expand and by 1973 they were producing over 750 million tons per annum, some 50 per cent more than the total production of the U.S.A. in the same year. Their increases in output were not, however, continuous and regular, as with the world's total oil production and consumption. Output from year to year in each country (see Map 4, pp. 90–91) depended upon the interaction of a series of factors, each of which influenced company decisions on how much oil should or could be lifted from a particular country or field.

The central element (though much modified by other factors, as we

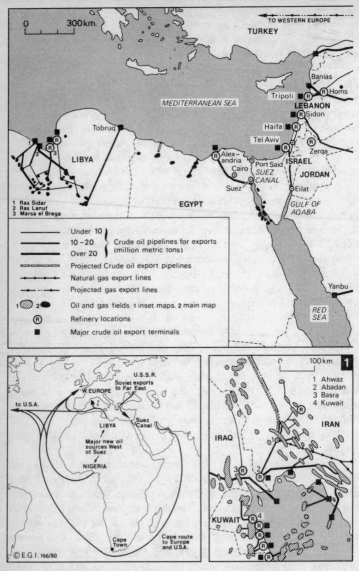

Map 4. The oil industry in the Middle East and North Africa.

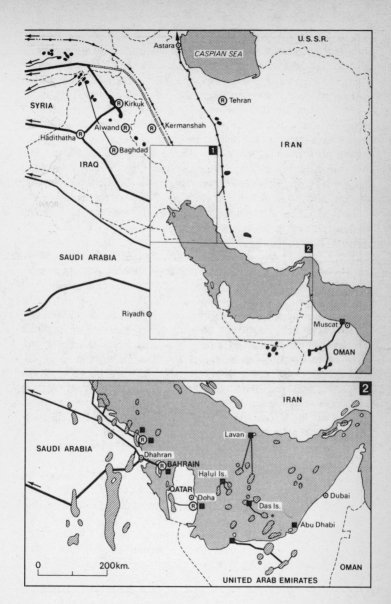

shall show later) in decision-making was the evaluation by the companies of the costs involved in producing and delivering oils of differing qualities to their various markets at any given moment, given the various permutations of producing and transport facilities available to them. Except in the early post-war period when, as already shown, the pressure of market demand enabled maximum output to be sustained from all producing areas, the amount of oil which could be produced in the Middle East through the facilities available has generally exceeded the amount of oil required at any time. Thus companies were able to make choices between the different countries according to contrasting costs and other factors. This was an advantage which the companies retained until their relationships with the producing countries were fundamentally changed in 1973, after which the countries themselves took over the right to make decisions on such important questions (this will be fully discussed later in the chapter).

But it would be a gross oversimplification to suggest that the consideration of comparative costs of production and transport in the short term was the only, or even the main, factor involved in determining individual companies' choices of patterns of production. Production in each of the four main producing countries – Iran, Iraq, Kuwait and Saudi Arabia – was controlled not by individual companies, but by a consortium of two or more companies. Each company thus had some influence in determining oil output in each country, the changing fortunes of which, however, depended upon the compromise reached between all the companies concerned after discussion and negotiation. In that the joint preferences undoubtedly differed from individual ones, the pattern of output from the Middle East countries was certainly not the same as it would have been in the case of an uninhibited search for lowest-cost production opportunities by each producing company acting independently.

Another main constraint on the automatic choice of the lowest-cost output pattern was the reaction by the companies to the political situation in the different countries at different times. The political stability of the early post-war period in the Middle East soon gave way to grave instability arising from the growth of nationalist feeling, the decline of Britain's influence and the conflict over the establishment of Israel. Thus, particular governments from time to time either took action which so upset the oil companies that they restricted their production and development efforts, or they took measures which made it impossible for the companies to continue to operate normally.

An outstanding event of this kind occurred in Iran in 1951. Between

1945 and 1950 conditions for expanding output were so favourable and so much new capital was invested there by the Anglo-Iranian Oil Company that production increased from 16 million to 32 million tons per year. Then, in 1951, the industry was nationalized by the then Prime Minister, Musaddiq. Nationalization not only stopped development, but also virtually stopped production too. Indeed, until the issue was resolved by negotiation three years later the production of oil from the Iranian fields was limited to the million or so tons required annually for domestic consumption. The nationalized Iranian oil industry lacked the ability to sell abroad the oil produced from freely and willingly negotiated concessions made previously between government and company and so recognized by customary international law. The Anglo-Iranian Oil Company's threat to take legal action against any entity 'buying' and using its oil was sufficient at that time to ensure that there were very few customers for it, and thus virtually no Iranian oil moved on to world markets over this period.

By 1954 the oil that could have been produced from Iran was flowing from additional producing capacity that had been established in Kuwait, Iraq and Saudi Arabia. Agreement between the oil companies and Iran was finally reached in 1954 and a new type of relationship between government and companies was established. The former retained ownership of the oil produced – as well as the oil reserves remaining in the ground – but it permitted the companies to continue to exploit the fields in the former concessions of the Anglo-Iranian Company and made the companies responsible for selling the oil abroad as contractors to the National Iranian Oil Company. The agreement brought in companies other than the Anglo-Iranian, which had previously had sole concessionary rights, and oil quickly started to flow again from the existing fields. In addition, the new consortium of contracting companies moved in to explore for fields in the other parts of the country where they were permitted to work under the terms of the agreement. This arrangement was highly successful and both old and new fields repeatedly established production records. In 1971 overall production topped 200 million tons for the first time. This period of rapid development of Iran's oil industry was clearly due to the acceptability of the 1954 agreement to the companies and to the fact that Iran enjoyed a relatively high degree of political stability in the second half of the 1950s and in the 1960s, when it also remained unaffected by the troubles with Israel. This, of course, was in marked contrast to the situation during this period for the Arab oil-producing states of the Middle East. These political advantages for Iran, coupled with the low cost of much of

Iran's oil production and its proximity to the coast and the new off-shore tanker terminals that gave a low transport cost component in the free-on-board cost at point of export, enabled the country to achieve the highest rates of oil production growth amongst the Middle Eastern countries in the 1960s.

But even rates of increase in output averaging nearly 15 per cent per annum – nearly twice the rate of growth in world demand for oil – were not always accepted as satisfactory by the Iranian government, and by the latter part of 1966 there appeared to be a danger of a major upheaval in government/company relationships. The Iranians sought to establish the principle that what the oil companies did in various countries by way of exploration, development and production should be related to the development needs of those countries, particularly in terms of the total populations that had to be sustained and provided for out of oil revenues. Iran, with over 30 million people at that time, claimed that a 15 per cent rate of growth in annual oil production was too low to meet the needs of the country for development capital and that the companies were discriminating against Iran by producing too much oil from countries (such as Kuwait and Saudi Arabia) with small populations and limited needs for revenues.

Thus, for 1967 the government insisted that the rate of growth in oil production should be at least 17 per cent, a rate of increase which it claimed was physically possible given the known characteristics of the oilfields. The companies concerned did not seriously challenge the validity of this claim, for their searches had been enormously, and somewhat unexpectedly, successful. If they agreed, however, to increase their rate of offtake from Iran to this degree, then they faced trouble elsewhere, for it would mean reducing the rate of growth in output from other producing areas in the Middle East. The other countries concerned, with their much smaller populations, could not be expected to see eye to eye with Iran's proposal that oil offtake should be related directly to the development needs of the local populations. The oil companies also calculated that to concentrate their oil development efforts on Iran would be more expensive overall than the alternative production strategies open to them in their several producing areas in the Middle East. At first, therefore, they fought Iran's proposal with a great deal of vigour, but ultimately they were more or less obliged to acquiesce, under the threat of what amounted to the cancellation of their contractual rights to work in Iran and the prospect of 'their' oil being sold in world markets through the assistance of the Soviet Union and of companies in the western world not associated with the Iranian producing consortium. The changed government/company relationship arising out of the 1954 agreement between Iran and the Consortium had removed the legal right of the companies concerned to claim ownership of the oil from their contractual areas in Iran when sold on world markets. They were now only contractors working under licence from the Iranian national company, and thus the outcome of the dispute was in marked contrast to the outcome of the 1951 'nationalization' dispute which, as indicated above, closed down production. This change gave a first clear indication of the rising significance and power of the producing nations in the post-war oil industry, not simply in terms of their ability to win an increasing share of the profits, but in terms of their influence on decisions on levels of production and their rates of increase – this important extension of their power was to be fully achieved within a few years.

Similarly, in other parts of the Middle East the 'oil fortunes' of individual countries have varied according to changing political circumstances. In Iraq, changes in government on several occasions in the 1950s and 1960s caused the Iraq Petroleum Company (a consortium in which five of the international majors had an interest) to cut back from time to time on production levels and/or investment plans, with the result that oil output increased in a series of fits and starts. In 1965–6 greatly increased port dues on Iraqi oil shipped out via the Shatt al Arab made the oil cost more than alternative supplies from other parts of the Middle East, with the result that the southern Iraq fields were partly closed down. Moreover, the country's northern fields have had their output affected from time to time by other difficulties over transport, arising in this case from circumstances beyond the control of the Iraqi government. This is because production from Iraq's northern fields depended for its outlet to markets on crude oil pipelines that had been built through other countries to the Mediterranean coast (see Map 4, pp. 90–91).

In the early post-war period a problem arose because these lines passed through territory which, in 1948, came to form part of the state of Israel. Arab opposition to the formation of Israel meant that these lines could no longer be used, and northern Iraqi oil production could only be resumed after the completion of new arms of these lines to new export terminals in Lebanon. The new lines, however, also passed through Syria, whose extremism in the Suez crisis of 1956–7 caused their sabotage and closure, so that Iraq's northern fields again had to be shut for a period of several months. Then, in the mid-1960s, oil production from the fields ceased for a third time as a result of disagreement between the

Syrian government and the pipeline company over the royalties to be paid by the latter for the right to transport the oil across Syrian territory. In December 1966, after a long period of unsuccessful bargaining, the Syrian government formally took over the pipeline installations as a means of bringing pressure to bear on the company to increase its payments. The Iraq Petroleum Company reacted, however, by diverting tankers normally calling at the Mediterranean coast terminals to other loading points. Thus, the storage capacity at Banias and Homs was quickly filled, the flow of oil through the pipelines was brought to an end, and oil production in the northern Iraq fields had to be stopped. Agreement on the royalty transit rates was eventually reached in March 1967, but in the meantime Iraq lost the royalties and other revenues which would have accrued from four months' oil production, with consequent damage to its economy.

In light of these repeated difficulties and the continuing uncertainty over the future prospects for these pipelines through Syria to Lebanon, the Iraqi government then took steps to establish other transport routes for oil from its northern fields – albeit somewhat more expensive ones – to serve markets west of Suez. The first was a pipeline direct to an Iraqi port on the Shatt al Arab at the head of the Gulf and the second was a pipeline from the oil fields to a terminal on the eastern Mediterranean coast of Turkey, thus avoiding Syrian territory. The completion of these alternative transport facilities (see Map 4, pp. 90–91) eliminated one cause of the fluctuations in Iraq's production.

However, even while these new transport facilities were in the planning and the construction phases there were other problems which affected Iraq's production. These arose from difficulties in the relationships between the Iraqi government and the Iraq Petroleum Company, which, with its wholly owned subsidiaries, was then still responsible for Iraq's total oil production. This dispute seriously affected output for some years after 1967, when the Iraqi government unilaterally took back about 99 per cent of the acreage of the country conceded to the I.P.C. for petroleum exploration and development. It did this because it claimed the need to exploit as quickly as possible the oil resources in these areas which, the government argued, the I.P.C. had effectively sterilized, as its member companies preferred to make their exploration efforts in other parts of the world. The I.P.C. denied this allegation and claimed that it was giving as much attention to developing Iraq's resources as the prospects would allow. The dispute continued through to 1971 and made the I.P.C. do little more than mark time in its production schedules. In the meantime, however, the government offered some of the former I.P.C.

acreage to other companies, and it also made arrangements for some of the potential reserves to be explored and exploited by the Iraq National Oil Company. The arrangement involved the help of the Soviet Union which, in return for loans at low rates of interest and for technical help, was to be reimbursed out of any eventual crude oil production.

Only the unexpected agreement in 1973 between Iraq and the I.P.C., following the general agreement on a *modus vivendi* between the oil-exporting countries and the major international companies in the period after 1971, prevented what could otherwise have been a battle royal between the parties concerned. In this battle Iraq would almost certainly have been supported actively by the forces of Arab nationalism, by other petroleum-exporting countries, by the Soviet Union and probably also by some commercial as well as political interests in the west (for instance, the French and Italian state oil entities backed by their respective governments) anxious to see the power of the international majors curbed still further. The I.P.C. for its part could normally have expected the unflinching support of the U.S., British and Dutch governments but this was not forthcoming. There were several reasons for this. First, because there was some obvious validity in the Iraqi claims that the I.P.C. had not chosen fully to explore the areas allocated to it and, moreover, as a result of the dispute, had chosen to limit Iraqi oil production; second, and more important, because of the changed geo-political conditions in the Middle East in particular, and the oil world in general. Thus they threw their influence behind getting the I.P.C. to concede the very considerable amount of ground needed to reach an agreement acceptable to Iraq. This the I.P.C. finally did, but thereafter it still remained to be seen whether I.P.C.'s reaction to its defeat would be to put Iraq low on its list of countries for investment in oil exploration and production facilities. On the other side, what now had to be demonstrated was whether or not Iraq's new partners, especially the Soviet Union, in the acreage given up by the I.P.C. would be willing and/or able to help in developing new production facilities as quickly as Iraq might wish. These questions were, however, soon overtaken by events. The new-style arrangements between the oil-producing nations and the oil companies, following the late 1973/early 1974 establishment of the former's unilateral rights to take whatever decisions they liked in the ownership, production and price of oil without reference to the oil companies, made a solution to the Iraq/I.P.C. dispute irrelevant.

In 1974 the Iraqi petroleum industry was completely nationalized. Since then the state oil company has attempted to expand both production and exploration activities, with contracted help from both private and public foreign companies and institutions. Its efforts were successful and by 1979, when Iran's production was greatly reduced consequent upon the revolution, Iraq became the second largest oil producer in O.P.E.C. (after Saudi Arabia). Its level of output increased from just over 2 million to almost 4 million barrels per day and Iraq was reluctant to constrain production, relative to the potential which the I.P.C. had earlier been unwilling to develop. Politics, however, intervened once again to upset the planned expansion of Iraq's oil industry. Iraq went to war with Iran in 1980 and this has led to the destruction of, or damage to, the infrastructure needed to export oil, particularly its terminals and storage facilities in the Shatt al Arab which, given the problems, as described above, of Iraq's export facilities to the Mediterranean, had become relatively more important in the post-1971 expansion of production. Iraq's oil production in 1980 declined substantially (to average less than 3 million barrels per day over the whole year) but then in 1981 there was a dramatic decline to under 1 million barrels per day. Annual production since then has been of the same magnitude and with the war with Iran still continuing the prospects for higher levels of output are not bright, in spite of the fact that Iraq is developing new transport routes for its exports – through pipelines to Saudi Arabia's Red Sea coast. Even if the war ends a return to the production levels of 1979 is unlikely given the weak conditions of international demand that have emerged since then and the more intense competition between supplying countries.

In Kuwait and Saudi Arabia the growth of oil production had, until recently, suffered less serious fluctuations. Moreover, those there have been have arisen from quite different circumstances. Both countries have been capable of producing much more oil than they have in fact produced in most years since 1950. Just how much more they could have produced is, however, difficult to estimate because the companies operating in Kuwait and Saudi Arabia never had any obligation to publish figures on their production capacity or their reserves. Generally, therefore, the companies responsible for oil production in these countries have been able to respond readily and immediately to reduction in supplies from other exporting countries. For example, when Iranian exports ceased in 1951 Kuwaiti production was stepped up considerably over a period of a few weeks in order to make up the total supplies required by the world market. Later, in the second half of the 1960s, Kuwait and Saudi Arabia 'suffered' from the rise in Libyan output, which increased to 150 million tons by 1969, for without this development in a low-cost producing country much nearer to the markets of Western Europe, they would have been responsible for much of the tonnage produced in Libya. In part also, their changing levels of production have depended upon the markets achieved by the companies responsible for producing their oil. This has been especially significant in the case of Kuwait, where until the early 1970s all production was the responsibility of the Kuwait Oil Company, in which only B.P. and Gulf Oil had interests. The total market outlets of these two companies were always too limited to enable them to exploit Kuwait's production potential to the full, even after they secured additional outlets for part of their Kuwaiti production through long-term sales contracts with other major companies (for example Gulf Oil with Shell) and even after seeking, very aggressively, the contracts to provide crude oil to independent refiners in many different parts of the world.

Nevertheless, in spite of these factors both Kuwait and Saudi Arabia enjoyed continually rising production levels for the whole of the post-war period from 1946 to 1972. It is not a coincidence that this continuity of growth occurred only in the oil-exporting countries in which there was an absence of radical political change. They were, moreover, countries in which external major western powers with oil interests, viz. Britain and the U.S.A., retained considerable political control (in the case of the former) or influence (in the case of the latter). Neither Kuwait nor Saudi Arabia was, however, content with this situation. Thus they both sought greater independence of Britain and the U.S.A. and the elimination of the monopoly control exercised by a single foreign owned and directed consortium over the oil sector of their economies. Both thus formed local oil companies to participate in the refining, transport and local marketing of oil, and these state-owned entities gradually built up their experience, expertise and investments in chemical plant, tankers, etc., in order to participate more widely in the industry. Both countries, following negotiations with the consortiums concerned, also limited the areas under concession to them. Some of the areas relinquished were then offered to other parties for exploration and development so that oil production in each country could become the responsibility of several groups, rather than just one. This, of course, would reduce the risks to the maintenance of production levels. Moreover, the new agreements, such as that reached between Kuwait and Japan in 1968, usually carried stringent requirements concerning the size and speed of the exploration and development efforts which the concessionary company was to undertake. This was in contrast with the freedom that the consortia had enjoyed to determine for themselves the speed and the extent of their commitments to exploration and development. They also obliged the companies concerned to maintain appropriate production levels once oil

99

reserves had been established. This measure was initially designed to overcome the problem that Kuwait had already experienced with B.P. and Gulf, in the period when they 'sat on' discovered resources to the detriment of Kuwait's revenues, but more recently the provision was also used to enable the government of Kuwait to put an upper limit on the amount of oil produced by the companies. This was done in the context of the country's interest after 1971 to create an upward pressure on the price of oil and its need to reassure itself about the ability of its discovered resources to sustain adequate production levels in the long-term.

Since 1974, of course, these revised arrangements for oil concessionary rights of the late 1960s and early 1970s in Kuwait and Saudi Arabia have been overtaken by events in the oil world. Kuwait, like Venezuela, has nationalized its industry and it also has negotiated technical, managerial and commercial arrangements with the previous concessionary companies. This means that it is now the decisions of the Kuwait National Oil Company which determine the speed and the intensity of new developments, as well as levels of production from existing facilities. Saudi Arabia took a 40 per cent interest in the consortium responsible for its oil industry, viz, the Arabian–American Oil Company (Aramco), in 1975 and its national oil corporation is now responsible for decisions on extending the oil industry to parts of the country in which no development to date has taken place. An agreement in principle, whereby Saudi Arabia was to take over Aramco completely, was reached in early 1977. It was finally implemented in September 1980. The delay arose from continuing negotiations over more general U.S./Saudi Arabian relationships by means of which the U.S. sought arrangements to secure a guaranteed steadily increasing supply of Saudi Arabian oil, in order to meet its expected oil import needs. Agreement along these lines seems to have been achieved so that the level of Saudi Arabian oil production was tied in to a special relationship with the United States. In the context of this agreement, the expertise and knowledge of Aramco, the consortium of American companies responsible for making Saudi Arabia the world's largest oil-exporting country (and the country with the largest potential for much increased near-future production), remained on hand – and was adequately, even if not handsomely, reimbursed – to ensure the continued flow of oil. The first results of the arrangements made became apparent following the outbreak of the Iraq–Iran war in late 1980. Saudi Arabia very quickly announced its decision to increase its production of oil as much as necessary to avoid shortages arising from the cessation of Iraqi exports. More recently, however, the overall decline in the demand for oil in general, and for O.P.E.C. oil in particular, has led to a quite dramatic cut-back in Saudi Arabian oil production. From a level of ± 10 million barrels per day, in the months after Iranian and Iraqi production was cut back because of the war between them, its rate of production fell back to ± 6 million barrels per day by the second half of 1982. For the first time in the history of its oil industry Saudi Arabia was not able to sell as much oil as it would have liked, and the special arrangement with the United States, which is importing more Mexican and North Sea oil in spite of an overall reduction of its import needs, came under severe pressure, particularly when the United States ceased to require more oil for its strategic stock-pile. Since 1983, moreover, Saudi Arabia has had to bear the brunt of the severe cut-back caused by the quite dramatic fall in the demand for O.P.E.C. oil. By mid-1985 Saudi production had declined to little more than 2·2 million barrels per day (110 million tons per year) compared with almost 6 million b/d only three years earlier.[1]

Venezuela and the four countries of the Middle East already considered in this chapter have produced almost 90 per cent of all the oil traded internationally in the period since 1945 and thus quite obviously dominate in any examination of the interests of the producing countries. But two further aspects of these interests must also be surveyed: first, the interests of minor and/or newer producers; and second, the significance of the formation of an oil producers' association – the Organization of Petroleum Exporting Countries, or O.P.E.C., as it came to be known.

One noticeable feature of the post-war period was the disappearance as 'oil powers' of countries which had considerable influence as producers and exporters up to 1939. Mexico was, until recent oil developments there, one example. For a short time in the early 1920s Mexico was the world's leading oil producer outside the U.S.A. and it continued as an exporting nation of some significance (contributing about 10 per cent of oil entering world trade) right up to 1938. Nationalization of the oil industry in that year, however, cut off exports immediately, as the dispossessed companies threatened legal action against anyone 'purchasing' oil from the new state entity, Petróleos Mexicanos (PEMEX). (This was identical to the action taken over the nationalization of Iranian oil in 1951, see p. 93.) The differences between the Mexican government and the oil companies were eventually resolved when the former agreed to pay, and the latter agreed to receive, compensation for the nationalized assets over a period of twenty-five years. This made it possible for

1. The issue of Saudi Arabia's role in the international oil market over the period since 1973 is discussed at greater length in Chapter 9.

Mexican oil to enter world markets free of legal restraints, but even then exports were not resumed on anything more than a minor and intermittent scale for many years. Until recently PEMEX had more than enough problems to cope with in ensuring that it could meet the rapidly growing oil and gas requirements of the Mexican national market. Moreover, in the 1960s it was prevented from offering crude oil for sale overseas by a decision of the Mexican government which argued that the country should itself secure the economic advantages to be gained from refining the unprocessed raw material into more 'valuable' products. Unfortunately, the markets of the world were not very anxious to accept Mexican oil products in preference to supplies of crude oil from elsewhere, and thus no significant export trade in products could be built up. Mexico thus became a self-sufficient, rather than an oil-exporting, nation, except for a little border trade with the U.S.A. (Mexican oil was not subject to U.S. quotas), and for small-scale and intermittent sales to importers in Western Europe. Post-1975 discoveries of large new oil and gas fields in southern Mexico, however, quickly changed this situation. These fields were developed rapidly and the additional production soon generated an export surplus. By 1980 Mexico exported over 50 million tons and by the end of 1982 oil exports were running at the annual rate of almost 90 million tons, in spite of the weak international oil market. Though Mexico has since agreed to moderate its export growth rate in response to O.P.E.C.'s pleas for it not to undermine international oil prices, further increases in Mexican oil exports may be expected over the next few years. Indeed, Mexico not only seems set to become one of the U.S.A.'s main sources of oil and gas, but also an important exporter to Japan, Western Europe and Latin America. (See Chapter 9 for a discussion of the implications of this development.)

On the other side of the world, Indonesia – formerly the Dutch East Indies – was an important pre-war supplier of oil, not only to markets in Asia and the Far East, but even to Western Europe. The Japanese occupation and the destruction of oil-producing and other facilities during the war eliminated this Dutch colony as an oil-exporting region in the early post-war period. Re-investment by Royal Dutch/Shell, the company most concerned, immediately after the war following the re-establishment of Dutch rule, soon reactivated the oil fields and the export terminals, etc., and a production of some 10 million tons was achieved in 1948. Most of this went to Japan, Australia and other relatively close markets to which Far East oil had a transport cost advantage over Middle East supplies. But the possibility of increasing production above that level was soon thwarted by fundamental disagreements between

Royal Dutch/Shell and the newly independent Indonesian government. Long negotiations and various attempts to establish new arrangements for exploration and development which would not only satisfy the national aspirations of the newly independent country, but also be acceptable to the company, were unsuccessful. Thus, little additional output was achieved and until the late 1960s Indonesia remained a small contributor to international trade in oil – even within the Far East, whose markets came, in the main, to be supplied from the Middle East. Since then, however, Japan has taken an increasing interest in Indonesia with a view to reducing its dependence on the Middle East, and important new agreements have been concluded for Japanese exploration of Indonesian oil areas (see also Chapter 6). The results of this work were very promising – so much so, in fact, that companies from other countries have also sought and secured acreage, particularly in Indonesia's offshore waters, and it is now certain that this country's steadily expanding role since 1970 in providing oil and liquefied natural gas in international trade will expand very markedly over the rest of the century. Its proximity to the world's largest and still expanding markets for imported oil and liquefied natural gas – in Japan and the newly industrializing countries of the Pacific Rim – has acted as a powerful stimulus for development, particularly since the governments indicated their intentions to diversify the region's oil and energy imports away from its more than 80 per cent dependence on the Middle East (see Chapter 6). These intentions arose out of the oil supply crisis in late 1973, following the Arab cutbacks in oil exports as a reaction to the war with Israel, and have been confirmed by a series of subsequent disruptions of supply and other difficulties. Indonesia has thus been climbing back towards its earlier relatively more important role as an oil producer and exporter and this is set to continue as increasing numbers of foreign oil companies respond to the relatively attractive terms offered for oil and gas exploration and exploitation by the Indonesian government. The government has a motive for pursuing such a policy given the country's rapidly growing population (already over 120 million). It thus needs to seek higher revenues by increasing its oil and gas exports. In this respect it is unlike some other member countries of O.P.E.C. which, with their very small populations, have very limited needs for rising revenues.

Another Far Eastern country, Burma, was before the war an oil-exporting nation of some importance, but now it is a country producing only just enough for its still very limited domestic use of oil. Oil industry development there was also prevented by difficulties between the government and companies which formerly operated the concessions.

Now the Burmese oil industry, like the Mexican, is wholly within state hands, though with some help from foreign companies as contractors, and it is only since 1978 that there has been any interest in Burma in the revival of an oil export trade. Foreign companies have now been invited to participate in oil industry developments and a number of new discoveries have been made. It is, however, still too early to judge the potential results.

Thus, there have been and, indeed, still are some developing countries which consider the ideology of national control over their oil resources as more important than the opportunity to expand oil production and exports by allowing foreign companies suitable concessionary agreements. Other countries in the Third World, on the other hand, saw the opportunities provided by oil developments for enhancing government revenues and solving balance of payments problems as more important in determining their attitudes towards foreign oil companies. Most significant amongst the countries with this kind of attitude were two in Africa, viz. Nigeria and Libya.

Exploration for oil in Nigeria started as long ago as 1938, but wartime disruption and then a greater interest immediately after the war in other parts of the world by the two companies concerned, Shell and B.P., prevented serious attention being paid to Nigeria until the late 1950s. Exports from Nigeria were initiated as late as 1962, but from then until the outbreak of the civil war between the Federal government and the breakaway state of Biafra – in whose territory many of the oilfields were located – production and exports went ahead very rapidly. This was due in part to the fact that Shell and B.P. were spurred to greater activity in the late 1950s than they had shown previously during the period when they monopolized the Nigerian industry. This was a result of concessions being granted to other oil companies whose initial exploration efforts proved to be very successful, in spite of the great physical difficulties (climatic and physiographic) of looking for oil in the Nigerian coastal region, and they quickly initiated new developments which soon led to increased production and exports. By 1966 Nigeria's exports exceeded 25 million tons per year and the plans and projects by the various companies then under way indicated a possible export level of 50 million tons by 1970 or 1971. This rapid expansion of Nigerian oil took place under the stimulus of a very favourable petroleum law which, amongst its provisions regulating oil development, included one with the effect of limiting the percentage of profits taken by the government to only about 35 per cent. This figure was little more than half that taken at the time by the major exporting nations (a situation which changed when Nigerian tax rates were raised after 1970 to the general level set by member countries of O.P.E.C., which Nigeria joined that year). Those lower tax conditions certainly encouraged companies to press on quickly with their developments, but equally favourable in this respect was Nigeria's location with respect to the Western European and North American markets. Oil exports from Nigeria did not have to go through the Suez Canal and there was, in any case, a much shorter haul from Nigeria to the main markets of Western Europe and the United States than from the Middle East (see Map 4, pp. 90–91).

Unhappily, further progress in the development of the oil industry in Nigeria after 1966 was temporarily restrained by the civil war. Indeed, even the existing industry was badly affected so that exports in 1968 were virtually nil. With the end of the civil war and the re-establishment of order, however, Nigeria's advantages as an oil exporter quickly led to the renewal of the strong upward trend in the country's contribution to world oil trade. This development was already apparent by the end of 1969 following the Federal forces' recapture of the main oil-producing areas and export terminals. Thereafter the restoration of peace in 1970 quickly led to a full resumption of oil activities in what had been Biafra, and this was soon followed by the successful development of new oilfields well to the west of the areas affected by the war. This enabled 53 million tons to be produced from Nigeria in 1970. Production increased to over 70 million tons in 1971 and to over 100 million tons in 1974, when Nigeria became the world's fifth largest oil-exporting country. It was, of course, at that time unhindered by the 1973–4 limitations on exports engendered for the main producing countries of the Arab world by the Arab–Israeli war. Overall, oil company efforts in Nigeria were greatly stimulated after 1970 by the country's favourable location for European and North American markets (see bottom left inset, Map 4, pp. 90–91) and in the later 1970s by Nigeria's non-involvement with the political problems of the Middle East, as well as by the high quality of its oil, with its low sulphur content and its high yield of the more valuable lighter products such as gasoline and diesel oil. Production in 1979 was, as a result, even higher than the previous peak in 1974. Since 1980, however, Nigeria's oil production and exports have suffered more than oil from other O.P.E.C. countries from increasingly serious competition from British and Norwegian North Sea oil, much of which is of a similar high quality and all of which has been available to refiners in both Western Europe and North America at a lower price.

Nigeria's earlier period of rapid development in the late 1950s and the early 1960s was, as indicated above, influenced by a favourable petroleum code. The same happened in Libya. Exploration there dates only from 1961, when oil industry interest in North Africa was stimulated by major oil and gas discoveries by French state enterprise in neighbouring Algeria. Thereafter, Libya's favourable petroleum law attracted many companies to join in the search. At this time the costs of possible failure of exploration efforts in Libya were not inflated by large payments which had to be made to the government irrespective of what the exploration produced, as was the case for new concessions in some older producing countries. As things turned out, however, many of the eagerly participating companies in oil exploration in Libya quickly achieved success, to such a degree, in fact, that exports from Libya began in 1962, only eighteen months or so after the first success of the initial exploration effort. By 1969 exports had reached a level of 150 million tons, putting the country into third position in the world league table of petroleum-exporting countries at that time.

The annual rates of increase in production in Libya over this early period of its development greatly exceeded those of other producing nations, as companies took advantage of the favourable taxes. The introduction, however, of a new, comprehensive oil law in 1966 appeared to threaten this situation. From the point of view of the companies the most disadvantageous provision of the law was Libya's claim to a share of the profits from production equal to the share claimed by the 'traditional' oil-producing countries. As this had the instant effect of reducing the companies' profits by at least one-third, some of the companies re-thought their expansion programmes given the ready availability of additional oil-producing potential in other countries. They thus turned back, in part at least, to other exporting countries for the incremental tonnage they required to serve their expanding markets in the mid-1960s. There was thus the prospect that the rate of increase in production in Libya would fall away, although the prospect of continuing growth was not in doubt because additional companies were achieving production for the first time and pipelines and terminals were being completed to move their oil away to world markets. Indeed, one of the main reasons for Libya's rapid development as an oil-producing country lay in the multi-company nature of the process of oil exploitation in Libya. No fewer than eight groups of companies, embracing fourteen companies in all, quickly achieved production or made significant discoveries. This sort of multi-faceted development of the country's oil industry was in marked contrast with the single company or single consortium development in other major Middle Eastern producing countries where, as shown above, developments had only been possible at the speed acceptable to the least enthusiastic member of a consortium.

However, just as the established producers in Libya were preparing to consolidate their positions in the country, rather than planning for further rapid expansion, they were presented with the clearest possible demonstration of the important geographical advantage which Libya has, even more than Nigeria, over the Middle Eastern countries in respect of getting its oil to the markets of Europe and North America. The advantage of the shorter distance from Libya to these markets was already considerable, but to this there was now added the further advantage of the country's location on the 'market side' of a closed Suez Canal. When the Suez Canal was blocked as a result of the Middle East war in June 1967 the companies abandoned their plans for merely consolidating their positions in Libya. They turned, instead, to an alternative strategy of rapid expansion of their production capacity in Libya, in order to achieve the savings in transport costs arising from the higher productivity of tanker use in getting Libyan oil to markets. Shipping oil from the Mediterranean to Western Europe and North America saved at least half the transport costs compared with the journey round the Cape of Good Hope from the Persian Gulf (see Map 4, pp. 90–91). Thus Libyan oil was eagerly sought and the country's physical facilities were rapidly expanded to make greatly increased production and export possible.

In 1968 the companies achieved no less than a 50 per cent increase in production and exports over 1967. In 1969 rapid development continued so that production and pipeline capacities were increased by almost another 50 per cent. In the circumstances of the continued closure of the Suez Canal, Libya could, within a year or two, have become the world's third largest oil producer (behind only the U.S.A. and the U.S.S.R.) and the largest oil exporter, overtaking even Venezuela, which had up to then held that position since the end of the war, and Iran, whose production was, as shown previously, being increased more quickly than in most other countries as a result of its need for development funds. This development, however, required that Libya itself remained interested in its production being rapidly increased. This, however, soon proved not to be the case as the country's oil policy underwent a fundamental change under the new, radical rule of Colonel Qaddafi who came to power by a successful coup against the western imposed and supported régime of King Idris in early 1970.

Thus, after the middle of 1970, the new Libyan régime joined Venezuela in arguing for higher oil prices, and it limited some companies' output in order to bring the necessary pressure to bear. As Libya, with its small population, had little immediate need for additional revenues from oil exports it was able steadily to build up pressure against the companies, with a consequential near-stabilizing effect on the country's total production. Then, in 1973, the prospect for a reduced level of production emerged from the nationalization of some of the country's important oil-producing operations, especially as there were no immediate plans for alternative ways of exploiting and selling the oil hitherto produced and marketed by the expropriated companies. Thus Libya had already embarked on a course of action which meant the severe limitation of its production well before 1973, when production limitations became a general policy aim of the Arab oil-exporting nations in the aftermath of the October war with Israel. Libyan production in 1974 was little more than 50 per cent of that in the peak production year of 1970. This was a much more severe cut-back in output than that for any other member country of O.P.E.C. and it reflected both the greater willingness of the Libyan government to use oil as a weapon in the war against Israel, and also the very much more radical policy which was immediately implemented by Libya against the oil companies in the country. As indicated above, the process of nationalization had already begun early in 1973 before the more general oil supply crisis of later that year. In the aftermath of the supply crisis, the process in Libya was accelerated.

It took time for some of the companies to accept the loss of their assets as a *fait accompli* and they tried to make do without Libyan oil, so extending the period of severely reduced production from the country. At a later stage agreements between the government and some of the companies on compensation for the nationalization of the industry were reached and these enabled sales and marketing arrangements to be implemented. Thus, production and exports were able to recover from their low 1974–5 levels in 1976 and 1977 but then they stagnated under the generally weak international market conditions. Production has since never recovered to the 1973 level of about 110 million tons, let alone to the much higher levels of production and exports achieved in the early 1970s. This, in part, has reflected Libya's lack of interest in maximizing its oil output. At the higher prices at which oil could be sold in the mid- and late 1970s it had no need to pursue such a policy and, indeed, it even threatened at the time of the supply difficulties in world markets in 1978–9 (at the time of the Iranian crisis) to stop exporting oil for two or three years. In part, however, the reduced level of Libyan production also reflected the companies' – and many Western importing countries' – recognition of the more radical and uncertain Libyan policies over oil in particular, and over relationships with the western world in general.

108

These acted as constraints on the willingness of both western companies and governments to commit themselves to too much Libyan oil in spite of its quality (low sulphur) and locational (closer to markets) advantages. In addition, since mid-1981 Libya has also been adversely affected by falling international demand for oil in general and, as with Nigeria, by competition from similar North Sea oil. Production in 1981 was over one-third down on 1980 and the low level of output has since persisted – at under 50 million tons – partly because of the weak international market but also specifically as a result of a U.S. ban on imports of Libyan oil as a reprisal for Libyan attacks on U.S. interests in the country. There still seem to be no politico-economic factors which indicate the likelihood of a major change in this situation for some time.

Finally, in this chapter concerned with the policies and fortunes of the oil-exporting countries, we must examine the significance of the formation and success of the body these countries created to represent and protect their interests in the world of oil and international politico-economic relationships, the Organization of Petroleum Exporting Countries (O.P.E.C.). Unlike nations concerned with the production of other primary goods, the world's major oil-exporting countries enjoyed, throughout the post-war period, the benefits of a rapid and continuing increase in the demand for the commodity they produced. Until 1973 they did not suffer from the severe fluctuations in demand and price levels that have, for example, afflicted the countries that produce metals, such as copper and tin, or agricultural raw materials, such as cotton and wool. They therefore became used to the idea of each new year being better than the previous one in respect of the level of oil industry activities and the flow of government revenues from the industry. As shown in Chapter 1, the existence of a small number of very large international companies working informally together through the late 1940s and most of the 1950s to control most of the world's markets for oil guaranteed this situation. By 1959 these companies were, however, under some pressure in the market place from each other as well as from outsiders, and they thus decided to reduce their tax commitments to the producing countries by reducing the prices they posted for crude oil. For most oil-exporting countries at that time those posted prices determined the level of the revenue payments which the companies made to the governments. This decision caused an unexpectedly strong reaction in the oil-producing countries concerned. The very short-term result was that the companies had to restore some of the price cuts. An almost equally quick response was the formation of O.P.E.C., as the main Middle Eastern oil-producing countries were finally persuaded by Venezuela and Libya that such an international oil producers' organization was essential to curb the freedom of action of the international companies in their dealings with them.

As has been explained earlier in the chapter, Venezuela had very strong economic motivations to secure such a collective agreement, while Libya's new régime wanted to constrain the power of the companies as a matter of principle. The positive reaction of the Middle Eastern countries towards the proposal for such an organization was at this time essentially political, for their reserves and production potential were large enough to ensure rising total revenues even with reduced revenues per barrel. However, the validity and importance of such political motivations should not be understated. Indeed, it was in this political direction, rather than in an economic sense, that O.P.E.C. was mainly significant until 1971. Politically, it served to demonstrate the potential power of a group of developing nations against the international companies which had hitherto been able to 'play-off' any one of the countries against the others. It also gave the group of nations concerned much experience in working together in analysing problems which they shared, and so helped to prepare them for the subsequent battle of words and arguments against the oil-importing countries of the industrialized world.

In economic terms, however, O.P.E.C.'s success in the first eleven years of its existence was limited because, from the economic standpoint, the interests of its member nations differed too much. Thus, while they were collectively able to agree on and successfully work for the implementation of 'royalty expensing' (a technical change in the formula for calculating gross profits so that the countries' share was increased by a few percentage points), the organization failed entirely over the critically important economic issues such as the prorationing of output to agreed levels. Such control of output – in order to maintain prices – was really only in the interests of Venezuela, Indonesia and possibly Iran. For the others, the more oil they produced, from their apparently near-inexhaustible reserves, the better. However, it would in any case have been far too much to expect agreement quickly within O.P.E.C. on such fundamental economic issues. The most surprising thing about O.P.E.C. by 1970 was that it continued to exist ten years after its formation and that it was, indeed, becoming increasingly aware of the realities of world oil power and recognized as potentially important by the oil companies. Certainly the oil companies at first expected it to be nothing more than a very short-term phenomenon, but its continued presence through the 1960s at least persuaded them that they could no longer take unilateral decisions to reduce posted prices. They soon came to regard O.P.E.C. as having at least a nuisance value in hindering their efforts to organize the world oil industry – but later they came to recognize it as a means whereby oil prices could be maintained and possibly increased.

O.P.E.C. thus gradually came to represent an important element in the growing sophistication of the oil-producing world. Then, with its collective decision made in Caracas, Venezuela, in December 1970 to pursue policies to assume control over the supply and price of oil and, meanwhile, to present a joint demand to the producing companies for increased royalties and taxes on oil, O.P.E.C. not only secured an immediate significant increase in oil prices, but also demonstrated clearly that the balance of power in the oil world was moving away from the oil companies, and in favour of the nations with significant oil resources on which most of the rest of the non-communist world had increasingly come to depend for its growing energy needs. This success of O.P.E.C. clearly demonstrated to its members the advantages of collective action and it then sought to operate as a kind of producers' cartel. By mid-1973, after complex negotiations with the oil companies (dealt with in more detail in Chapter 9), it had already secured two further sets of increases in posted prices and hence in government revenues. Even more important, from a longer-term point of view, it had secured oil company recognition of the validity of producing-country 'participation' in the producing operations. It was thus quite clear by the very early 1970s that the international oil companies had had to recognize the strength and significance of O.P.E.C. as a new power in the world of oil. So much so, in fact, that the companies now determined that their own immediate best interests (that is, the maximization of their short-term levels of profitability) lay in their cooperating with the oil-producing countries against the interests of the oil-importing countries. O.P.E.C.'s collective action in raising revenues from oil could, they found, be used as a convenient argument to justify significant price increases for oil products in the consuming countries. This cooperation between the producing countries – working through O.P.E.C. – and the international companies, which had also been working jointly together for some years since 1968 through the London Oil Policy Group, was a development of immense importance in the distribution of world oil power. The immediate impact of such cooperation was, indeed, strongly felt in oil price rises and oil shortages in many consuming countries between 1971 and 1973 – in the two years, that is, before the first oil 'crisis'.

At this stage O.P.E.C. may well have become counter-productive for the oil-producing nations' best interests, because its success in cooperating with the already somewhat suspiciously viewed international oil companies was beginning to indicate to the oil-consuming nations that a 'plot' existed to increase the cost of their essential oil imports. This could well have led to the formation of an O.P.I.C. – an Organization of Petroleum Importing Countries: a development which had hitherto proved to be unnecessary in the period when O.P.E.C. failed to have any significant effect on the price of crude oil. The new inability, and unwillingness, of the oil companies to absorb the additional costs created by the success of O.P.E.C., coupled with the post-1970 inability of the consuming nations to take any further measures to reduce the cost of their oil imports (until 1970 such measures had been possible and will be discussed in later chapters), seemed as though they might generate the motivation and the collective will for action against O.P.E.C. and the international oil companies. Such action, had it been taken in the 1971–3 period, could have produced some constraint on the degree to which increasing revenues and profits were achieved, by the oil-exporting countries and the international oil companies respectively, from the supply of cheaply produced oil to the industrialized countries – and to the poor, developing countries.

Many of the oil-importing and consuming nations, including Japan and Western Europe, as well as some of the nations of the Third World in Latin America, Africa and Asia, were at that time already beginning to make alternative provision for their energy needs, particularly by the local development of oil and gas and by the more rapid expansion of nuclear power. Given a few years, the expansion of these energy resources, coupled with serious efforts to achieve a reduction in energy use, would have diminished the rate of increase in oil output in the major producing and exporting countries. Had the rate of increase thus been forced below the hitherto 'accepted' level, then the producing nations would have had to face up to the prospect of lower earnings from oil exports, and so been confronted with the same kinds of economic problems that all other countries outside the industrial world had had to face since 1945. One or more of the oil exporters could well have then decided to seek higher earnings by increased sales at lower prices, and have thus broken the collective strategy of O.P.E.C. and the international oil companies before it really got going.

This possible scenario for a collective oil-importing nation response to undermine the rising power of O.P.E.C. was, however, not followed. The most important consuming countries failed to note the change in 1970–71 in the world oil power structure, and they also chose to ignore the need to constrain the demand for oil in order to take the pressure off the constrained supply. Added to this, the prospects for a massive increase in U.S. imports as a consequence of the ending of the oil import quota system by the Nixon administration (see Chapter 1) was presented as a net addition to the volumes of oil exports required from the O.P.E.C. countries to sustain the increase in oil imports which were said to be needed by Western Europe and Japan. In addition to this powerful set of economic circumstances, in which potential demand for O.P.E.C. oil clearly threatened to outpace the supply of oil which O.P.E.C. members were prepared to make available (for the short to medium term), there were also the political circumstances of the latest Arab-Israeli war in 1973 which gave most of the O.P.E.C. countries – the Arab ones, that is – an immediate incentive to cut back on their deliveries of oil to a pro-Israeli Western world.

Out of this economically and politically engendered short-term scarcity of oil, monitored and indeed controlled by O.P.E.C. and supported by the oil companies, there arose the opportunity for the price of oil to be raised to whatever levels the exporting countries chose to ask. The consequence was the traumatic fourfold price increase in the posted prices of crude oil between October and December 1973. This was accompanied by what turned out to be the even more significant notification by the O.P.E.C. countries that they intended to disregard the agreements they had reached with the oil companies in 1971–3 in respect of all essential decisions over the supply and price of oil. Furthermore, they also indicated that they would take over the oil companies' operations in the producing countries in a very much shorter time than had previously been agreed with the companies. These few months in the latter part of 1973, at the end of a period of three years from December 1970 during which O.P.E.C. had mapped out its future strategy, took the world into a fundamentally new position as far as the distribution of oil power was concerned. O.P.E.C. and the oil-producing nations now assumed a large degree of control in the international oil system. This fundamental change in the world of oil power from 1970 to 1975 is discussed in Chapter 9.

5. Oil Policies in Western Europe

Before the Second World War the economy of Western Europe remained very largely dependent on coal as its source of primary energy, with very little of the diversification into oil and gas that had already occurred in the U.S.A. (see Chapter 2). Between 1939 and 1945, however, the European coal industry was badly hit by wartime dislocation and destruction and much of the productive capacity in countries such as West Germany, Belgium and France was put out of action because of difficulties both in the mining areas themselves and in associated transport facilities. Even in places which were not directly affected by land fighting or intense bombing – most significantly, Britain's coal-mining areas – the coal-mining industries were, nevertheless, affected by the running-down of facilities, the lack of capital investment and difficulties in obtaining labour. Thus, in 1946 coal output in Western Europe totalled only 340 million tons (with Britain contributing nearly 60 per cent) and the prospects for a rapid growth of production were anything but bright. Most Western European countries were thus brought face to face with a probable serious deficiency in the total energy supply required for their post-war reconstruction and development.

In such a situation the possibility of obtaining oil supplies from overseas appeared to offer a sure and an almost immediate solution to the impending energy crisis, and increasing quantities of petroleum products started to move to Western Europe from refineries in the U.S.A., the Caribbean and the Middle East. In this early post-war period, oil imports had to be largely in the form of finished products rather than crude oil, as most of the limited amount of refining capacity that had been built in Europe before the war had, along with the coal mines and the railways, suffered physical damage and could thus only be brought back into production as rebuilding was completed.

Moreover, merely rebuilding was in any case only going to solve part of the problem. The refining capacity of Western Europe before the war had been small in spite of expansion in the 1930s, particularly in Germany and France, for strategic reasons. This was because the economic advantage prior to the war lay in refining oil at the source of the crude oil and in shipping the products to their markets in Europe. Not even relatively undamaged Britain had any refining capacity available to help meet Europe's oil needs. Indeed, it too had to import products from the U.S.A. and elsewhere.

The dependence of Europe on imported petroleum products – paid for in part by U.S. aid – did not, however, persist for more than a few years. First the larger countries, and then the smaller ones began to investigate the possibilities of curtailing the impact of growing oil products on their balance of payments. This arose because most of the oil available came either from dollar areas or through American companies which had to be paid in dollars, and hence produced a serious drain on Europe's very limited dollar resources at that time. These initial investigations concentrated on ways of encouraging the development of large oil refineries, so that the import of oil products could be substituted by the import of crude oil and its processing into products in Europe itself. This was a means of ensuring a reduction in the unit foreign exchange costs of Europe's oil requirements, even after allowing for the fact that much of the equipment required in the new oil refineries had to be bought in the United States.

Thus, European governments used both the carrot and the stick to persuade the companies which supplied the oil to implement this policy. The companies were, for example, given loans at less than the then current rates of interest for financing the construction of refineries and they were also guaranteed the availability of foreign exchange to import the necessary equipment. At the same time, however, they were told that once the refineries had been established, then preference would be given, through tariff or quota arrangements, to crude oil imports over product imports. This, of course, meant that companies which failed to build refineries could well find themselves forced or priced out of the markets.

As things turned out, however, the companies really needed little or no encouragement to accept official policy. Development in transport and refining technology, coupled with the rapid growth in Western European demands for oil, soon persuaded them that the establishment of refineries in Western Europe offered a lower-cost and more profitable way of meeting these demands than the traditional pattern of importing oil products. The growth in the size of crude oil tankers brought down the unit cost of crude oil transportation very considerably, whilst the rising volume of demand in the industrialized parts of Western Europe meant that larger refineries could be constructed, thus enabling significant economies of scale to be achieved in the new projects. As oil was imported to provide energy to markets in those areas of Europe which had formerly depended almost exclusively upon coal, the rise in demand for fuel oil outstripped that of all other products. In fact, it increased sufficiently to ensure the local use of all the fuel oil that the new refineries could make. This contrasted markedly with the pre-war situation when the demand for fuel oil had been so small (it had been unable to compete with indigenous coal) that even Western Europe's small refineries had been obliged to re-export some of this product. This had meant higher transport costs on the whole operation, so making it more expensive than the alternative of manufacturing the products near the sources of the crude oil.

Thus, from the points of view of both Western European governments and the large international oil companies there were, in the years immediately after the war, strong economic motives for the construction of large oil refineries. By 1950 the first of the post-war refineries or major refining expansions – for example, at Fawley in the U.K., Pernis in the Netherlands and Marseilles in France – were already successfully operating, mainly on crude oil imported from the rapidly expanding fields of the Middle East (see Chapter 4). Very shortly afterwards the economic arguments which favoured refinery construction in Europe were reinforced by strategic and political ones. Before the war, the major oil-producing areas had been securely under British and French control, but in the aftermath of the war the Middle East entered on a period of extreme nationalism during which the influence of Britain and France declined very markedly (this is described and explained in Chapter 8). The consequent political instability, both of the region in general, and of individual oil-producing countries in particular, persuaded both the importing nations and the large international companies that it would be unwise to make additional investment in refineries in the Middle East. It became clear to both governments and companies that it would be better to limit investment in the oil-exporting areas to that which was required for the oil-producing operations, and, in order to spread the risk, to build the refineries instead in countries where the degree of political stability was greater, and where the likelihood of expropriation or of disruption to operations was less than in the Middle East.

Various events quickly demonstrated the validity of the political and strategic considerations involved in this policy. These included the closure of the pipeline from Iraq to the Mediterranean at Haifa and the consequent closure of the refinery there when, as a result of the formation of the state of Israel in 1948, the Arab nations refused to continue to make crude oil available to it. Shortly afterwards, in 1951, the Anglo-Iranian Oil Company was nationalized and the country's oil-producing facilities and the world's largest refinery at Abadan were closed down for more than three years. As oil became more important in the economies of the Western European countries, so the degree of risk that they were prepared to accept over oil supplies became smaller. Thus there was an escalating interest in the development of local refineries and, by the mid-1950s, Western Europe had sufficient refinery capacity to meet its main oil products needs (though minor products still had to be imported). The continent's main estuaries and other favourable locations became dotted with refineries (as shown on Map 5, pp. 126–7) and the pattern of Western European trade in oil had now changed from one in which the main imports were oil products to one in which the crude oil requirements for the continent's new refineries were imported. By this time the annual saving of foreign exchange achieved by the development of a European refining industry was of the order of $150 million.

Western Europe was not, however, satisfied merely with the successful establishment of a refining industry in respect of its growing dependence on oil from overseas for its economic rehabilitation and growth. There was also concern over the international companies' controlled system of oil pricing, for this also gave Western Europe an unnecessarily high burden of foreign exchange costs for its growing oil requirements. In the early post-war period there was still an element of influence on prices from the international oil cartel which had been established by the major oil companies in the early 1930s as a result of the world depression. The cartel had instituted a pricing system which kept prices higher than they need have been in most markets. The system had also restricted the development of low-cost sources of oil as a result of its aim to equalize the delivered price of oil, no matter what its source. Thus, higher-cost oil from the United States (for example) continued to be exported to secure markets in Western Europe in spite of the availability of much lower-cost oil from the Middle East.

The system worked in roughly the following way. Posted prices for crude oil and oil products were established in U.S. Gulf ports and freight rates were declared for the transportation of oil from these export ports to the import ports of Western Europe. Oil from the inherently lower-cost producing nations of Venezuela and the Middle East was only made available by the international companies at the same prices, on a grade-for-grade basis, as the Gulf of Mexico posted prices. The delivered price of this oil in Western Europe was, moreover, then calculated by the companies concerned as though it was shipped from the Gulf of Mexico, rather than at a price reflecting the shipping costs of the actual oil involved. Thus the prices of oil in Western Europe were not only tied to prices which reflected the relatively high costs of the regulated oil industry within the U.S.A. (so that the much lower costs of production in the Middle East or in Venezuela, whence most of Europe's supplies originated, were not taken into account), but they also reflected higher than necessary ocean transport charges.

Thus, in the late 1940s Western Europe was paying, on average, over $4 for each barrel of crude oil imported and even higher prices for imports of oil products in a situation in which oil was available from the Middle East at well under $1 per barrel. The governments of Western Europe were naturally not at all happy with this situation. In 1950, following pressure by European governments as well as by the U.S. authorities responsible for the Marshall Plan (by means of which some of Europe's oil imports were being financed), the first break in the pricing system was forced on the companies. They agreed to post separate prices for the oil which they produced in Venezuela and in the Middle East. Though these posted prices were at first not much lower than those of the equivalent grades at the Gulf ports of the U.S.A., they quickly became so as price levels in the U.S.A. rose in the post-war inflationary situation. The effect of this was, of course, a more favourable situation for the Western European oil-importing nations. In addition, separate freight rates reflecting the use of larger, lower-cost tankers were established for direct movements of oil to Europe from the Caribbean and from the Middle East. As a result of these important changes the average delivered price of oil per barrel in Western Europe came down to under $3·50.

Even this general price level was, however, well above the supply price of oil – that is, the lowest price at which the company with the highest costs for oil exploration and production had to sell its oil in order to achieve an adequate rate of return on its investments, including due allowance for the risks involved in the enterprise. The continued existence of such a favourable price for crude oil sold to Western Europe made the oil industry a highly profitable one and this encouraged the entry of new producers into the industry in many parts of the world. Many potential producers looked particularly to the marketing opportunities in Western Europe where, under the combined impact of rapidly expanding economies and the increasing substitution of oil for coal, the consumption of oil increased for some years at a rate of about 15 per cent per annum. New companies – both American and European – started to ship oil to Western Europe, with the result that the posted-price system and the assessed freight-rate system, which together had fixed an 'appropriate' price for the delivery of any grade of oil to any port in Europe, came under competitive pressures. The systems eventually broke down as the following developments occurred.

Marketing companies and refineries in Western Europe which were not owned by, or committed to take their oil supplies from, the major international oil companies started to shop around for the crude oil and/or oil products they required, and increasingly found them at prices well below those quoted by the major companies on the basis of posted crude oil prices and agreed freight rates. At first the activities of such companies affected only the fringes of most of the national markets, but as their success started to take business away from the international companies the latter were forced to respond in order to keep their own local companies competitive and profit-making. They were thus obliged to lower the prices at which they transferred oil to their affiliated refining and marketing companies in most European countries, with the result that practically the whole of the market, and not just the fringe business, started to enjoy lower prices. These naturally led to an even greater growth in oil consumption so that in the period between the late 1950s and 1970 the use of oil in Europe went ahead at a more rapid rate than ever before. This expansion in the use of oil meant that coal made a decreasingly important contribution to Western Europe's increasing energy needs. Indeed, oil replaced coal as the most important source of Western Europe's energy in 1966 and by 1970 had achieved this position in all but one Western European country; the exception was Britain, whose coal industry managed to hold on just a little longer against competition from oil before it too had to accept second place in the national energy supply situation. This process of change from dependence on coal to dependence on oil in a selection of Western European countries is shown in Table 1 (pp. 120–21).

Thus, in the main, Western European countries achieved their immediate post-war objectives of increasing the size of their oil-refining industries, and of securing their oil imports at prices somewhat more closely related to the costs of production in the areas from which most of Europe's oil originated – the Middle East and, somewhat later, North Africa. There have, of course, been country-to-country variations in the implementation of these policy objectives. Italy, Scandinavia and West Germany pushed the policies to their ultimate conclusion, whereas Britain, the Netherlands and France tended to have somewhat different objectives as a result of their own particular interests in other aspects of the international oil industry. France, for example, gave preference to oil from the franc zone, notably Algeria, and maintained prices at a level which made this possible. It also decided to encourage the establishment

Table 1. Energy use in selected Western European countries, 1937–85 (in million tons coal equivalent – MTCE)

Year	Energy Use	United Kingdom	West Germany	France	Italy	Netherlands	Belgium/ Luxemburg
1937	total MTCE	202	142	82	20	15	35
	tons per capita	4·3	3.0	2·1	0·7	1·8	4·0
	% coal	74	97	89	70	87	97
	% oil	26	2	8	18	13	3
	% natural gas	–	–	–	–	–	–
	% other*	–	1	3	12	–	–
1952	total MTCE	232	145	89	25	22	34
	tons per capita	4·6	2·9	2·1	0·6	2·1	3·8
	% coal	90	95	79	43	78	89
	% oil	10	4	18	35	22	11
	% natural gas	–	–	–	8	–	–
	% other*	–	1	3	15	–	–
1957	total MTCE	247	186	111	44	28	39
	tons per capita	4·8	3·5	2·5	0·9	2·5	4·2
	% coal	85	88	72	28	63	80
	% oil	15	11	25	48	36	20
	% natural gas	–	–	1	15	1	–
	% other*	–	1	3	9	–	–
1962	total MTCE	265	221	122	71	35	42
	tons per capita	4·9	3·6	2·6	1·4	2·6	4·4
	% coal	72	58	58	18	47	70
	% oil	28	33	33	62	51	30
	%natural gas	–	5	5	13	2	–
	% other*	–	4	4	18	–	–
1967	total MTCE	276	251	154	112	47	46
	tons per capita	5·0	4·2	3·1	2·1	3·7	4·7
	% coal	59	51	40	12	25	53
	% oil	39	46	50	71	58	45
	% natural gas	1	3	6	11	17	2
	% other*	1	1	4	6	–	–

Year	Energy Use	United Kingdom	West Germany	France	Italy	Netherlands	Belgium/ Luxemburg
1972	total MTCE	302	333	215	152	76	63
	tons per capita	5·4	5·4	4·2	2·8	5·7	7·1
	% coal	40	35	21	7	6	33
	% oil	<u>46</u>	<u>52</u>	<u>66</u>	<u>75</u>	36	<u>52</u>
	% natural gas	12	11	10	14	<u>59</u>	15
	% other*	2	2	4	4	–	1
1977	total MTCE	285	355	231	177	79	65
	tons per capita	5·1	5·8	4·4	3·1	5·7	7·0
	% coal	38	32	20	7	5	25
	% oil	<u>41</u>	<u>50</u>	<u>63</u>	<u>70</u>	36	<u>50</u>
	% natural gas	19	15	12	19	<u>58</u>	22
	% other*	2	3	5	5	–	3
1982	total MTCE	275	370	250	195	86	68
	tons per capita	5·0	6·0	4·7	3·3	5·9	7·0
	% coal	35	34	22	10	6	28
	% oil	<u>39</u>	<u>46</u>	<u>58</u>	<u>65</u>	40	<u>46</u>
	% natural gas	23	17	14	19	<u>52</u>	22
	% other*	3	3	6	6	1	4
1982	total MTCE	285	385	260	210	85	65
	tons per capita	5·0	6·1	4·8	3·4	5·8	6·3
	% coal	34	35	18	12	9	26
	% oil	<u>36</u>	<u>41</u>	<u>52</u>	<u>62</u>	42	<u>45</u>
	% natural gas	26	20	19	20	<u>47</u>	21
	% other*	3	4	10	6	1	8

Source: 1938–82, U.N. Statistics, Series J. Author's estimates for 1985

* Other – mainly hydro-electricity – and from 1972 also nuclear power – in MTCE on the basis of the heat value of the electricity produced.

For each year the most important energy source in each country is underlined. The percentages do not always add up to 100 because of rounding.

and development of French oil companies. The French government thus thought it necessary to exercise strict control over the oil industry in the post-war period, as indeed it was already doing before 1939, in order to ensure that national policies were respected by the international oil companies. Britain, on the other hand, was the headquarters of two of the international oil companies and it thus tended to consider that its national interest lay in the maintenance of oil prices at the international level so that those companies would not have their commercial interests adversely affected. This approach, coupled with the dominant 90 per cent share of the international companies in the British domestic market, ensured, at least until 1965, that the country's oil import prices remained well above the levels applying throughout most of the rest of Europe. In 1965 the average import price per barrel of oil into Britain remained some 25 per cent above the average import price into the rest of Western Europe. This was to the detriment of the country's balance of payments and it was also a continuing factor in helping to explain Britain's relatively poor economic performance compared with its neighbours. However, competition eventually came even to Britain and, as new companies moved in to take advantage of its highly profitable oil market, the established companies were obliged to shave their prices in order to meet it. By the late 1960s oil products' prices (excluding taxes) within Britain were not much higher than in most other European countries, although Britain still failed to enjoy the balance of payments advantage of the lowest possible import prices of crude oil to the same degree as Germany, Italy and Scandinavia.

There was another aspect of the energy situation in Western Europe that affected attitudes to oil – the impact of imported oil on the continent's indigenous energy supplies. This, of course, was of great importance for those countries where coal production was important. As we have already shown, the difficulties involved in increasing indigenous coal production in the early post-war period led to oil being welcomed enthusiastically as a means of overcoming the energy shortage. But the coal industries of Britain, West Germany, Belgium and, on a smaller scale, those of France and the Netherlands steadily overcame these difficulties. Recapitalization of the mines enabled mechanization and, later, automation to be introduced so that both production and labour productivity increased. As a result the continent's demand for coal was eventually met. By this time, however, in the later 1950s, as seen in Table 1, the demand for coal was already declining sharply as a result of its substitution by oil in many end-uses. By 1958 the pattern of post-war energy shortage – a pattern in which every ton of coal that could be won from the ground was more or less assured of a market at a price which was government controlled in order to prevent it rising too high – had changed into a situation of energy over-supply.

From then until 1973 the European coal industry suffered a period of more or less continuous decline occasioned by its inability to compete with oil in a wide variety of uses, from domestic home-heating to electric-power generation, and in basic industries such as iron, steel and cement, etc. Coal from the higher-cost mines of the continent became impossible to sell and thus every major coal-producing nation had a programme of mine closures, usually accompanied by measures designed to prevent social distress in the mining areas. But the programme of closures in almost every case moved too slowly to keep the relationship between supply and demand in balance, and other measures had to be taken in order to prevent the situation getting out of hand. Thus, most coal-producing nations were obliged to give protection to their coal industries, either through subsidies on coal production or by the imposition of taxes on oil products competing with coal in end-uses other than transport.

For example, in Britain a policy of rationalization was adopted by the National Coal Board, whereby many mines closed down and production was reduced from its post-war peak of 227 million tons in 1957 to only 130 million tons by 1973. This was nevertheless insufficient to produce equilibrium between the oil and coal industries. In the early 1960s the government was obliged to put a tax of 0·83p per gallon on oil products used for heating purposes. This tax – since raised by stages to over 3p – may seem to be relatively insignificant compared with the tax of 27p on each gallon of gasoline but even so it represented at the time, when oil prices were so low, a purchase tax on the ex-refinery price of fuel oil of up to 50 per cent. According to estimates made by the Ministry of Power in the middle 1960s this tax on fuel oil and other heating oils maintained the output of coal at a level some 18 million tons per annum higher than it would otherwise have been. Such taxation measures were common to all the major coal-producing nations of Western Europe. In addition there were other measures which were designed to force 'captive customers', such as government departments and other state-owned entities, to burn coal even when oil was cheaper. All these steps did no more, however, than ameliorate the increasingly adverse situation for coal, the economic viability of which was seriously undermined by the growth of oil.

At the beginning of the 1970s no more than 80 to 100 million tons of Britain's coal output of about 150 million tons remained competitive with oil without protection. This situation for indigenous coal was even worse in the rest of the continent. The virtual closure of the French and the Dutch coal industries was in prospect and the loss of over two-thirds of the coal industry of Belgium and some 60 per cent of that of West Germany seemed likely. And this was in spite of governmental measures designed to help coal in the Western European energy market-place. By 1972 oil provided almost two-thirds of Europe's total consumption of energy and it was no longer correct to describe Western Europe as having a coal-based economy. Its economy had developed into one based on the availability of two competing sources of fuel and power, with the advantage lying strongly in favour of oil throughout most of the region.

During the post-war period many factors combined to produce a fundamentally changed geography of the oil industry within the continent. These included the much more widespread use of oil, both geographically and in terms of sectors, technical developments in refining and the inland transportation of oil, and changes in the political structure of Western Europe, such as the establishment of the European Common Market. Before the war the pattern of oil supply had been a simple one. Oil products were distributed in relatively small quantities by water, road and rail to the centres of inland demand from the coastally located oil-product import terminals or from the small oil refineries at ports such as Rotterdam, Hamburg and Marseilles. The immediate post-war development confirmed this pattern, but the oil by this time, of course, moved mainly from the expanded refining centres which were being established at convenient points for crude oil imports around the coasts of Europe (see Map 5, pp. 126–7). The quantities involved in some areas grew sufficiently quickly to justify the construction of product pipelines to take the oil from the refineries to the major internal cities. Important examples of this development were the oil product lines laid from refineries at the mouth of the Seine to Paris, from the Rotterdam refineries to the Ruhr and, in the U.K., the lines from refineries on the Thames and the Mersey to distribution centres in the Midlands.

At a later stage, however, it became clear that this method of supplying oil products did not represent the cheapest way of delivering oil to some inland centres, especially given the growing need for fuel oil which cannot, for technical reasons, be pipelined over long distances. So, various locations were investigated for the establishment of inland refining centres which could be supplied with crude oil through pipelines running from import terminals on the coasts, and from which local distribution could take place. As a result, refineries were built in the Ruhr and in the Rhine Valley and were fed by crude oil pipelines from the Rotterdam area and northern Germany. From the mid-1960s refining centres were also established in Bavaria, eastern France and Switzerland, dependent upon crude oil supplies coming in by pipeline from the southern European ports of Marseilles, Genoa and Trieste. The use of these Mediterranean ports as crude oil import terminals to serve the refineries in the centre of Western Europe did, of course, significantly reduce the amount of ocean transportation needed to provide Western Europe with its oil. The oil had only to be brought to these Mediterranean ports of southern Europe over the relatively short distances from the originating sources of the crude oil in the Persian Gulf/eastern Mediterranean and North Africa. In this pattern of transport the tankers were saved a return journey of over 7,500 kilometres around the coasts of Western Europe to the ports of northern France, the Netherlands and Germany (see Map 5, pp. 126–7). With the continued development in the size of the oil markets and in pipeline technology the possibility of serving even northern Europe (for example, the northern parts of Germany and Scandinavia) by pipelines coming up from the Mediterranean, rather than by tankers bringing the crude oil into nearby deep-water ports, was evaluated by a consortium of the most important oil-supplying companies. This possible development was, however, thwarted by three factors. First, there was the continued closure of the Suez Canal after 1967. The consequent enforced use of the Cape route around southern Africa meant that the Mediterranean ports lost much of their ocean transport cost advantage over ports in north-west Europe. Second, large quantities of oil from the North Sea fields of Britain and Norway were becoming available by the mid-1970s, and this greatly reduced northern Europe's need for oil from more distant sources. Third, after the steep oil price increases of 1973–4 the demand for oil ceased to grow at the high rates of earlier years and thus eliminated the anticipated need for new facilities.

Though the growth of the new patterns of Western European oil supply, processing and distribution has emerged largely from the developing technology of pipelining and refining, some of the change in the structure of the oil industry in Western Europe has undoubtedly arisen from the move towards economic and political integration through the establishment of the European Economic Community. The barriers against the movement of goods between the nations of the Common Market have been all but eliminated. In this situation the geo-political problems which might have arisen from the European-wide organization of the oil industry within a Western Europe of completely independent nations were greatly reduced. The international oil companies concerned viewed Western Europe as a single refining and marketing region and

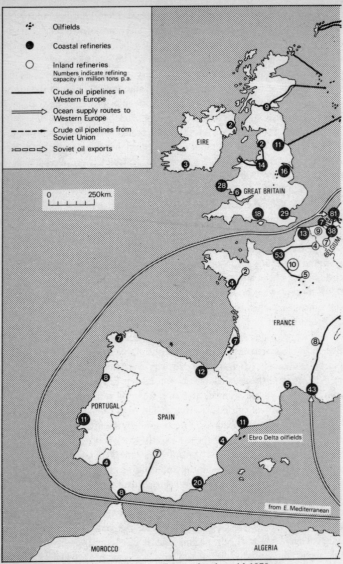

Map 5. The oil industry in Western Europe by the mid-1970s.

Note that much refining capacity has been closed or 'mothballed' since 1979 as a result of falling oil use and that most of the pipelines are

North Sea oilfields

NORWAY

FINLAND

SWEDEN

DENMARK

NETHERLANDS

WEST GERMANY

possible link W. BERLIN

EAST GERMANY

POLAND

U.S.S.R.

CZECHOSLOVAKIA

possible link between East and West European systems

AUSTRIA

SWITZERLAND

HUNGARY

RUMANIA

ITALY

YUGOSLAVIA

BULGARIA

ALBANIA

GREECE

TURKEY

from Libya

TUNISIA

© E.G.I. 167/80

also working at less than their designed throughput. Meanwhile, oil production, especially from the North Sea, has increased significantly.

they were thus able to plan and develop optimal refining and distribution patterns for the continent as a whole.

But this favourable impact of European economic integration on the oil industry, though important in contributing to lower costs for the industry and thus to the possibility of lower prices to consumers – which is the sort of favourable effect that customs unions are meant to have – has not provided the main topic for discussion between European oilmen and the politicians and administrators. At the time of the formation of the Common Market oil was still a relatively unimportant source of energy in the continent. The predecessor to the Common Market, the European Coal and Steel Community, was made responsible for the development of a European coal policy, but it was given no such authority over the oil industry. At that time the oil industry was considered to be of too little importance for the Western European economy to justify such governmental interest. Since the oil industry's rise to greater importance after the mid-1950s, however, many complicated and long-lasting discussions have taken place on the establishment of an overall energy policy, including oil, for the Common Market countries. However, because of the basic disagreements which arose between the interests of those members of the European Economic Community with indigenous coal production to protect and those without such interests, energy policy remained one of the unresolved and difficult sectors in the moves towards European economic integration. The need for an agreed energy policy for the whole of the area was apparent even in the period of decreasing real cost oil and energy, but it has been made more imperative by the fundamental changes in the oil price and supply situation since 1973. Efforts to achieve a common policy have thus continued but still without success. The differences, however, are no longer principally those between coal producers and the rest (in the aftermath of the oil crisis coal, as seen in the 1977 and 1982 data-sets in Table 1, was 'rescued' from its decline), but between the countries which want more control over the energy sector and those which will only accept less intervention. The emergent trend was, until very recently, in the direction of more effective and more comprehensive government control and, in this respect, French experience, dating back well over thirty years, perhaps gives the best idea of what such control implied.

A common energy policy for the E.E.C. would involve the establishment of some degree of preference for the indigenous energy industries which now, of course, include large-scale Western European oil and gas production. Whilst the oil industry would continue to be run by public and private individual companies it would have to be in the context of a basic framework which would be established by European civil servants. Such controls, which may seem to be anathema to the oil companies, would not be too unfamiliar to the U.S. oil companies as they are used to effective and comprehensive government intervention in their own country. Such a development would, moreover, produce a situation within which the companies could be certain of securing an adequate return on their investments, as it would take away much of the normal commercial risk to which they would otherwise be subjected. This could apply particularly in the case of 'Community' companies – that is, companies whose parent company 'belongs' to a member nation, as opposed to foreign-owned companies – for they would likely be accorded preferential treatment on both political and strategic grounds. Such a proposal was, indeed, made in 1982 by the E.E.C. Commission in respect of the steps necessary to rationalize the oil-refining industry.

The development of comprehensive energy policies in the Common Market and in other countries of Western Europe has in general been opposed by the oil companies. Until the events of 1973 they saw energy policies as the means of restricting their growth and freedom of action. Now, however, such policies could well become a means whereby the interests of the now declining oil industry – with an established position to defend in Western Europe's energy sector – are given a guaranteed role to play in the energy supply system. In the light of recent developments in the energy economy of Western Europe, in which growth in the use of energy has ceased and in which sources of energy other than oil are being encouraged, the oil companies' prospects and their future overall profitability could otherwise well be undermined.

The significant decline (by almost 25 per cent) in the use of oil in Western Europe since 1972 has already created problems of surplus capacity for the industry, particularly in respect of refining capacity. Projects which were already in development in 1973, when there was a general expectation that the use of oil would continue to grow quickly, were completed and thus created capacity which was not required as a result of the rapidly declining use of oil in Western Europe. Thus, many refineries have been closed but, even so, it will certainly be the end of the 1980s, or even much later, before use 'catches up' with the capacity which has been kept open or in 'mothballs'. In the meantime the companies are faced with costs which have to be recovered from a smaller than expected throughput.

In terms of energy supply the first development of significance which affected the hitherto rapid expansion of the oil industry and the growth in oil use was the discovery and production of large quantities of natural

129

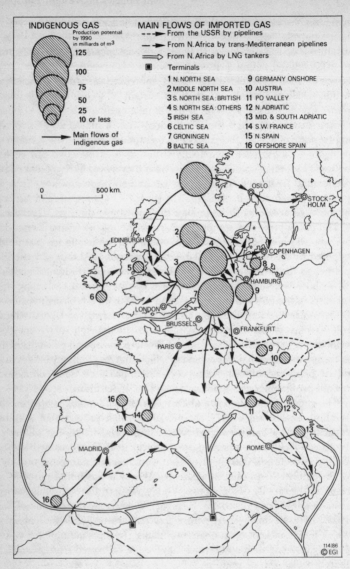

INDIGENOUS GAS
Production potential
by 1990
in milliards of m³

125
100
75
50
25
10 or less

→ Main flows of indigenous gas

MAIN FLOWS OF IMPORTED GAS
---→ From the USSR by pipelines
--→ From N. Africa by trans-Mediterranean pipelines
⇒ From N. Africa by LNG tankers
■ Terminals

1 N. NORTH SEA
2 MIDDLE NORTH SEA
3 S. NORTH SEA : BRITISH
4 S. NORTH SEA : OTHERS
5 IRISH SEA
6 CELTIC SEA
7 GRONINGEN
8 BALTIC SEA
9 GERMANY ONSHORE
10 AUSTRIA
11 PO VALLEY
12 N. ADRIATIC
13 MID. & SOUTH. ADRIATIC
14 S.W. FRANCE
15 N. SPAIN
16 OFFSHORE SPAIN

0   500 km.

OSLO
STOCK-HOLM
EDINBURGH
COPENHAGEN
HAMBURG
LONDON
BRUSSELS
FRANKFURT
PARIS
MADRID
ROME

114|86
©EGI

Map 6. Prospects for natural gas in Western Europe by 1990.

gas within Western Europe itself (see Map 6, p. 130). Natural gas had been available throughout most of the post-war period in northern Italy and southern France, but the quantities involved were relatively small and were capable only of providing a limited amount of the energy required in the two countries in a period in which energy demand was moving ahead very rapidly. The oil companies were little affected by this competition. The gas field of Groningen in the northernmost part of the Netherlands was, however, a very different matter. This field was discovered in 1959, but it was only slowly evaluated and declared. It is now known to be the world's largest exploitable gas field outside the Soviet Union and is capable of producing more than 100×10^9 m^3 of natural gas per year, representing the energy equivalent of over 80 million tons of oil per annum. As the premium or high-price markets for this gas in the Netherlands and nearby parts of France, Belgium and West Germany were satisfied, it was increasingly made available for use in processes and amongst consumers who turned to it in preference to using fuel oil, the markets for which were thus constrained after 1971.

While the most important source of Western European gas was confined to the Netherlands, however, the situation was never in danger of getting out of hand so far as competition with oil was concerned, for the Dutch gas field was very much in the hands of two of Europe's major oil companies, Shell and Esso. With the cooperation and agreement of the Dutch government, also a part-owner of the resources and also anxious to maximize its short-term gains from the gas, it was priced at a level for export sales that ensured it would not become quickly attractive to too many major oil consumers. The joint state/oil company enterprise – NAM – responsible for marketing the gas in foreign countries – France, Belgium, West Germany, Switzerland and Italy – thus ensured that its price at the Dutch border was high enough to limit its ability to compete against fuel oil, once transport and distribution costs to the foreign consumers had been added on.

This planned orderly marketing of Groningen gas, however, changed in the late 1960s. The change arose in the first place from the development of the North Sea basin's gas resources. When, as a consequence of the geological reinterpretation to which the discovery and importance of the Groningen field gave rise, it became evident that the whole of the North Sea basin was a potentially rich gas-bearing zone, the nations of Europe with shorelines on the North Sea met to discuss the way in which the ownership of these resources could be established and, as a result of their common interests, quickly reached an agreement on the division of the North Sea into a set of nationally controlled areas (see Map 7, p. 137).

Exploration of the resources lying beneath the bed of the North Sea has since been under way in the areas which were allocated to Britain, the Netherlands, West Germany, Denmark and Norway.

The first discovery of gas was made in British waters in 1964 and by 1969 the search under the British section of the North Sea had already been successful enough to determine a potential productive capacity equal to more than half of that of the Dutch Groningen fields. In 1975 Britain produced the energy equivalent of nearly 40 million tons of oil a year in the form of North Sea gas and by the end of 1977 had discovered sufficient reserves to enable production to rise to at least twice this level. In the case of Britain the quantity of gas available, the fact that it was being sold to customers by a national entity entirely outside the control of the oil companies, and the relatively low price at which the government ensured that it was sold by the producing companies to the British Gas Corporation, meant at first that natural gas replaced some of the fuel oil used by customers who had turned to the latter in preference to coal in an earlier part of the post-war period. This process of the substitution of oil by gas was enhanced by the large increase in the price of oil in 1973–4 and significantly slowed down the growth in oil consumption in Britain. Though the growth in demand for gas was then temporarily eliminated, partly as a result of industrial recession and partly because the Gas Corporation tended to overprice it as a deliberate policy to constrain growth, its use has more recently started to increase once again – partly because there have been significant additions to proven reserves as a result of continued exploration successes – and it now seems likely that the use of gas will provide up to 30 per cent of the country's total primary energy requirement by the end of the 1980s. Indeed, the continuation of the Dutch, British and, most recently, very large Norwegian successes in finding natural gas could well lead to the availability of enough gas in north-western Europe to provide over one-quarter of the total energy demand of the whole of this densely populated and highly industrialized part of the continent in the 1990s and, thereafter, well into the twenty-first century.

For more easterly and southerly parts of Western Europe there are other places from which this alternative source of primary energy is being made available in increasing quantities. This includes other indigenous gas from newly discovered fields in Italy and Spain and in several parts of their adjacent off-shore areas, but it also involves importing international gas on a very large scale from the known gas fields of northern Africa. This can be done both by liquefied natural gas tankers which bring gas from the massive fields of Algeria and Libya across to southern Italy and southern Spain and by the first trans-Mediterranean pipeline which has been laid from Algeria through Tunisia to Sicily and is capable of transporting up to 18 milliard m^3 of gas per year. A second line under the Mediterranean from North Africa to Spain is now under active consideration and seems likely to be built in the later 1980s. There are also prospects for shipping liquefied natural gas from Nigeria and other more distant producing areas though this involves expensive transport facilities. Meanwhile large-diameter pipelines running into Austria, Switzerland, West Germany and Italy have already been constructed from the gas fields of the U.S.S.R. and these facilities are in the process of being expanded so that even more Soviet gas can be imported by Western Europe over the rest of the century (see Chapter 3 for details of natural gas from the U.S.S.R.). There is, indeed, a prospect for a doubling of Soviet gas exports to Western Europe – to more than 70 milliard m^3. It is also possible that pipelines will eventually be constructed to bring natural gas to Western Europe directly from the fields of the Middle East. One project under discussion is for a line from Iran through Turkey into southern and, eventually, Western Europe, and gas from Qatar, one of the lower Gulf states with huge unutilized reserves, is also possible – and under discussion. Continuing developments in pipeline technology will help to make this very long-distance transmission of gas into the markets of Western Europe economically feasible. Such developments have already had the effect of making gas available in the consuming countries at prices which have enabled it to compete with oil products in an increasing number of markets, especially with the rapid and formidable increases in oil prices which Western Europe had to face after 1973.

This process will continue as gas supplies continue to build up from both indigenous and external producers. Western Europe's energy economy is, indeed, quickly becoming based on three fuels, as shown in Table 1 (pp. 120–21), in much the same way as in the U.S.A. and the U.S.S.R. The development is, as already indicated, partly at the expense of the existing markets for oil. At the expense, that is, of those markets which oil secured from Europe's other indigenous resource, coal, in the 1950s and the 1960s. This diversification has been welcome in every country of Western Europe where general concern has been increasingly expressed over the undue reliance on imported energy from politically unstable parts of the world. In spite of this concern, however, governments have not necessarily pursued appropriate policies to limit the degree of dependence on oil. Indeed, while oil remained as the only alternative to coal – an alternative which offered energy to industry, commerce and domestic consumers at prices well below those of domestic coal – there was little that Europe really wanted to do, or could afford to do, to reduce this dependence (except, as shown above, by ensuring the expansion of the refining industry at home, and the purchase of oil supplies from as many different overseas sources as possible). Natural gas introduced a new variable, for not only were indigenous supplies readily available and hence able to reduce the continent's high degree of dependence on energy imports, but it was also cheap to produce. This applied even when the comparison was with the low oil prices in Western Europe in the 1960s. More recently, of course, in comparison with the higher prices for imported oil which arose from the oil companies' and O.P.E.C. countries' agreements after 1971, and even more particularly from those resulting from O.P.E.C.'s control over the market since 1973, indigenous gas became, by comparison, very cheap indeed. Western European governments would thus seem to have had every possible incentive to ensure that the development of the continent's natural gas resources went ahead as rapidly as possible by pursuing policies which enabled the exploration and development work to be willingly undertaken by companies with long experience and the requisite expertise.

Such companies are, in the main, the oil companies we have already described as producers of oil and gas in other parts of the world. But elsewhere these companies have, at least until very recently, owned the energy resources which they produced from their operations and in return for which they paid royalties and other taxes to the governments concerned. They have thus been in a position to charge whatever prices the markets would bear for their products. Western European countries, on the other hand, have a long history of national control over the basic sectors of their economies and especially the public utilities such as gas and electricity. Given this economic philosophy, the oil companies have been generally restricted to producing the gas and have then been required to sell it to a single purchaser, viz. the national gas company of the country concerned. Much of the gas produced was, moreover, purchased from the companies on price and taxation conditions designed to give them no more than an adequate return on their investment and was then made available to consumers through public utilities at prices which could be related to the utilities' purchasing price of the gas, plus the costs of its transport and distribution, rather than to the price of oil in the market-place. In such circumstances, natural gas was able greatly to enlarge its markets, largely at the expense of imported oil. In recent years gas prices have moved closer to the equivalent of oil prices: this change in pricing policy has often left a price advantage for gas in markets when higher value oil products (such as gas oil and kerosene) have to be used. There thus remains an opportunity for gas to become an even more important element in Western Europe's energy economy though, it should be noted, this has been offset to some extent by the unwillingness of the main gas-producing countries, viz. the Netherlands, Britain and Norway, to produce as much gas as their potential allowed. Indeed, such institutional constraints have been the main barrier in recent years to the further expansion of Western Europe's natural gas production as the reserves have proved to be much more prolific than generally expected. However, under the stimulus of competition from external supplies of gas – notably the Soviet Union – these constraints are breaking down and, with a general consumer preference for gas over oil (and coal), gas markets in Western Europe now seem set to expand significantly over the rest of the century.

These continuing potentialities for an increased production and use of natural gas are more important than the better known, and much more controversial, continued large-scale development of nuclear power stations. Such development is, however, already under way in France, where at least one large atomic power station, of 1,200 MW or more, will be commissioned each year in the 1980s and where nuclear developments are intended to replace oil-fired thermal power stations. This is also the policy in Belgium. There will also be similar, though smaller scale, developments of nuclear power elsewhere in Western Europe. Thus nuclear power, too, will help to reduce oil's relative contribution to the region's energy economy. Indeed, the markets for some oil products have been declining even in absolute terms. This applies particularly in the case of fuel oil, against which imported coal has become highly competitive, especially for electricity generation. Imported coal has undercut the fuel oil price by 50 per cent or more, but even some indigenous coal has been able to compete especially as Western Europe's coal-producing countries – notably Britain and West Germany – have continued to give financial support to their coal industries for social and regional policy reasons. Fuel oil will also lose further outlets to nuclear power developments and will continue to be substituted by natural gas in many areas and end-uses. This process started in the Netherlands as long ago as 1968 when natural gas became the preferred fuel by most energy users. Between 1968 and the late 1970s the use of fuel oil in the Netherlands declined by about 50 per cent. In Western Europe as a whole fuel oil use in 1985 was only 50 per cent of the amount used in 1973.

Thus, total oil requirements in Western Europe seem likely to continue to decline slowly, or even rapidly as in periods of economic recession such as that of the period since 1979. Oil use will, however, continue to rise in some sectors – particularly motor transport – where there are as yet no generally available alternative fuels for gasoline and diesel engines – though even these presently 'protected' markets for oil products may not last for more than another decade as rapid advances are now being made in the technology of using other fuels. The consequential rapidly changing pattern of demand for oil products has necessitated changes in the continent's refineries in order to reduce the fuel-oil output which, until the mid-1970s, was the main product from most Western European refineries. Refining has thus become more closely tied in with petro-chemicals (whose feedstocks and fuel requirements can absorb the re-finery output for which other markets do not exist) and this has tended to move the preferred locations for refineries back to the coastal areas, where petro-chemical developments are usually more appropriate. This, moreover, has been in the broader context of a Western European re-fining industry which, as shown above, grew well beyond the size which demand required. Indeed, the oil-refining (and transportation/distribution) industry in Western Europe has already been cut back by about 30 per cent but, even so, it still remains much larger than it needs to be. This surplus will persist through the 1980s – to the discomfort of the oil companies faced with much increased unit costs of refining – unless additional steps are taken further to reduce capacity.

The companies themselves have, as indicated above, been doing just this – with the closure of refineries or reductions in capacity in most Western European countries – but in addition the E.E.C. also made an attempt to intervene in this sector of the oil industry since the suc-cession of closures started in 1978. Needless to say, the E.E.C.'s proposals raised issues for both governments and companies. These included questions as to whose refineries should be reduced in size or closed, and how compensation for such enforced retrenchment in the industry should be calculated and paid. Thus, nationalist as well as financial considerations arose and, not surprisingly, no agreement was reached at the E.E.C. level. The problem of excess refining capacity has, moreover, been yet further complicated by the growth in oil production in Western Europe itself (see below and Map 7, on pp. 137–9), as a result of which the producing countries, particularly Norway and the United Kingdom, thought to pursue policies which would encourage, or even require, the refining of larger quantities of crude oil at home. The ap-plication of such a policy by Norway (though not, in the event, by the U.K.) has, of course, led to the establishment of new resource-based refineries in an already overcrowded Western European situation. This development has made the problem for the companies of dealing with

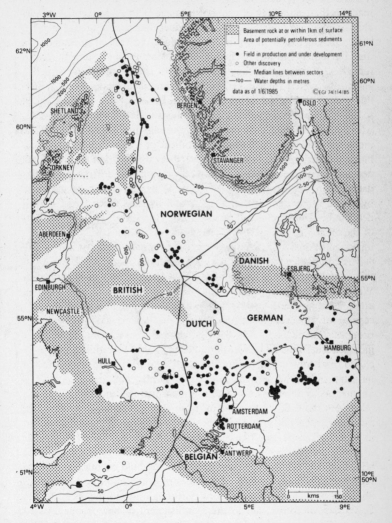

Map 7. North Sea oil and gas fields and discoveries to 1985
(see also next two pages).

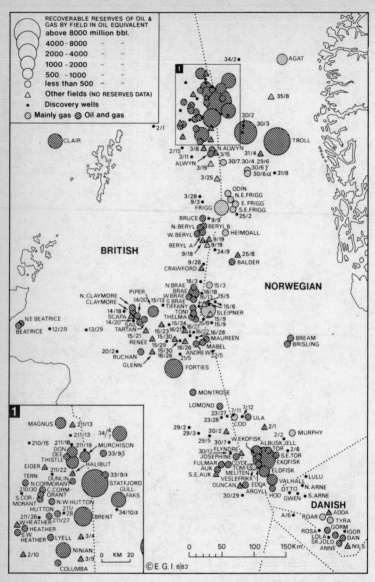

Map 7a. The north North Sea.

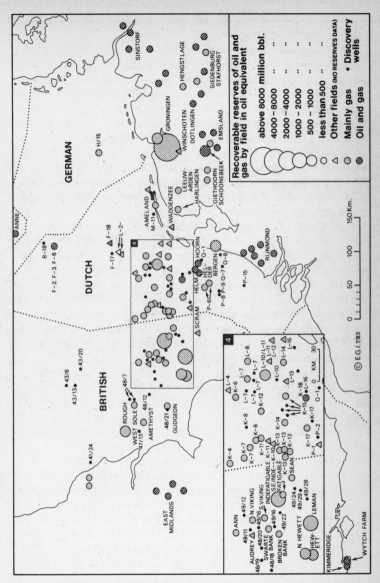

Map 7b. The south North Sea.

the now declining downstream sector of the continent's oil industry even more difficult to resolve.

These fundamental changes in the outlook for oil do, of course, have serious repercussions for the oil industry itself, but for Western Europe as a whole, its post-war security of oil supply problem – arising out of continuing difficulties in the Middle East – could by now have been but a spectre of the past had the challenge for producing indigenous oil on a large enough scale been accepted. It was not accepted, however, so that Western Europe continues to face a problem arising from its continued relatively high degree of dependence on imported oil, control over the supply of which has been exercised by the oil-exporting countries since 1973 rather than by the international oil companies whose efforts in the 1950s and the 1960s were responsible for making oil the dominant source of Western Europe's energy needs. Given the still important, though now declining, role of imported oil in Europe's energy economy, yet further interruptions of supply from one or more of the O.P.E.C. countries could still create difficulties in ensuring the supply of the oil needed, especially if demand were to move up sharply in the event of a recovery in Western Europe's economic fortunes. In the light of this one can, indeed, still argue the need for continuing efforts to curb the use of imported oil – by policies which deliberately seek to maximize the production of indigenous oil and/or to limit the volumes of imports – in order to eliminate the exporters' certainty of gain from supply interruptions; and thus to discourage such interruptions.

The achievements to date and the further promise of a greatly reduced dependence on imported oil has been opened up principally by the large and exciting new discoveries of oil resources under the North Sea. More than enough oil has been discovered to sustain the production of 200 million tons a year or even more from the Norwegian, British, Dutch and Danish sectors of the North Sea (see Map 7, pp. 137–9) over most of the rest of the century (production in 1985 was about 165 million tons). The development of the North Sea, moreover, marks only the beginning of the process of finding and exploiting Western Europe's offshore oil resources. Developments elsewhere to add to those in the North Sea could, by the mid-1990s, supply much of Western Europe's total oil requirements especially if in the interim period there remain effective restraints on the use of oil as a result of the combination of the very much higher prices which have had to be paid for O.P.E.C. oil (see Figure 1, p. 142) and the maintenance of effective oil-conservation measures.

Thus, after three decades of great concern over dependence on oil imports by European planners and governments, the continent should, by the late 1980s/early 1990s be able to afford a somewhat more relaxed attitude to the problem. This still depends, however, on policies being pursued which succeed in getting North Sea and other indigenous oil produced in the quantities which the resource base now makes possible. This is unfortunately still open to doubt as there are certain factors which effectively serve to restrain production rather than encourage it. We shall return to consider these issues in Chapter 10. Coupled with this possible lessening of fears about the security of oil supplies, there could also be some relaxation of the concern about rising oil price levels. Such concern became essential during the period of rapidly escalating prices after 1973 (see Figure 1, p. 142) but since 1982, given the pressure of competition both between oil supply sources and between oil products and other energy sources, oil prices in real terms have recently become significantly less unfavourable, as seen, that is, from the point of view of European governments concerned with oil-import costs, and of European consumers concerned with the cost of oil at the factory gate or delivered to a domestic storage tank.

Europe's 'battle for oil' could thus be all over bar the shouting, except for a continuing period of uncertainty during which the alternatives to imported oil will not be available in large enough quantities to enable the situation to be unequivocally controlled by the oil-importing countries. This still gives a remaining – though a declining – opportunity for O.P.E.C. countries to try to squeeze more money out of the European users of imported oil and thus it still presents a continuing potential danger to Western Europe. Nevertheless, given the constraints on O.P.E.C.'s freedom of action which have been introduced into the international oil system by overall energy policies (both national and those of the European Community), lower import prices should, as the rule, persist and even continue to weaken. Meanwhile, the companies concerned in the oil trade in Western Europe seem likely to have to be content with limited profit margins on the now more limited volume of sales which they can, at best, expect to achieve.

With this expectation of continuing relatively low profit margins on their European 'downstream' operations the companies concerned will have to devote increasing attention to the problems of achieving the lowest-cost continental scale supply and distribution patterns. A wrong choice by a particular company in this respect would easily eliminate the profit element entirely. The changes in the technology and the economics of refining and transportation and the changing geographical and product requirement patterns of oil demand further complicate the situation;

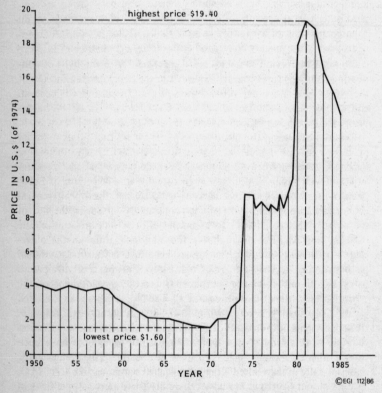

Figure 1. The price of oil, 1950–85 (shown in 1974 $ terms: in current $ the highest price – in 1981 – was $34 per barrel).

so much so, in fact, that we can expect to see even further developments in the already quite striking differences of opinion between companies about what they can best do to create the least-cost/highest-profit infrastructure in Western Europe. Previously, for example, one company chose to provide crude oil to its European refineries indirectly through a massive terminal in southern Ireland at which oil was received in 300,000 ton tankers and dispatched to various European countries in 50,000–100,000 tonners. Another company saw a trans-European (south to north) pipeline system, with refineries strung out along it in consuming centres, as the primary pattern of distribution at which to aim. Others saw the North Sea coast of West Germany, Britain, the Netherlands and Belgium as remaining the major growth area in the development of refining facilities.

As a result of the continuation of contrasting views by the different oil companies involved on the most efficient system to develop, the geography of the European oil industry seems likely to become even more complex in the period that lies ahead, particularly in the light of contrasting company (and country) interests in the development and use of the North Sea and other off-shore resources. These, indeed, pose an entirely new set of challenges and opportunities for infrastructural developments in the European oil industry over the rest of the century.

Finally, the organizational structure of the oil industry poses yet more problems. Its low degree of 'Europeanization' and the still increasing private U.S. companies' dominance of the European oil scene create a situation which may be decreasingly acceptable to a large part of European political opinion, particularly if U.S./Western European general economic and political relationships continue to worsen. In the rest of the century this could lead to enhanced efforts to 'Europeanize' the industry in a way not dissimilar to the 'nationalization' of the older-established energy industries in an earlier period of Europe's political history, and/or to the floating of many new European oil companies using locally-raised finance, particularly for the development of the North Sea and other off-shore resources. The Norsk Oljeselskap, floated on the Oslo stock market early in 1971 with an enthusiastic response from many tens of thousands of small investors in Norway, was the first example of this kind of development and since then there have been several other examples – such as Holland Sea Search in the Netherlands, and especially in the United Kingdom, where companies like LASMO, Tricentrol and Cluff Oil are now taking an active part as operators in North Sea operations.

Other routes towards ensuring a greater strength for local companies

143

on the oil scene include both that adopted in France, where a powerful state oil entity (E.R.A.P) has been established, and that in West Germany, where the government forced a set of small companies to merge to produce a viable oil company (V.E.B.A) able to operate at an international level, and in which the government itself took an important minority interest. Meanwhile the two countries most concerned with the North Sea developments – Norway and Britain – both created state oil companies (Statoil and B.N.O.C. respectively) to ensure that national interests in the development of their off-shore resources were increased. In both countries all new exploration and development work thereafter had to be on a joint venture basis in which the state companies had a 51 per cent interest. This strict requirement was relaxed in 1980 in the case of the U.K. as a prelude to the eventual 'privatization' of B.N.O.C. as part of the Conservative government's policy of reducing state ownership in the economy. B.N.O.C. became Britoil with 49 per cent private ownership in 1982 and the company was completely privatized in 1985, albeit with safeguards to limit the extent of non-British shareholding in the company. However, it still seems that an opportunity was lost in not creating entities which were specifically European in character and ownership as vehicles for developing Western Europe's own oil production and to give Europe an enhanced role in the international oil system. Such a European approach to the challenges and opportunities of the oil industry might well have been more effective in exploiting the continent's considerable oil wealth and in securing oil from other parts of the world. Such a development is now only possible as an integral part of overall moves towards more economic and political integration of the continent in the medium-term future. This is now less likely than it was, given the relative resurgence of nationalism in the last few years of increasing economic difficulties in Western Europe. Thus, the emergence of a 'European Peoples' Oil Company' now seems further away than ever.

6. Japan: Growth and Dependence on Oil Imports

By the end of the 1960s, apart from the few major oil-producing countries whose economies were then totally or largely dependent on the continuing production and export of crude oil, Japan appeared to stand as the country most susceptible to economic upheaval in the event of a major upset in the international oil industry. Japan's clear understanding, at the time of the oil crisis in 1973–4, that that, indeed, was the position was demonstrated by the speed with which it adjusted its policies towards Israel and the Arab States. The more favourable and supportive attitude it established in the immediate aftermath of the crisis towards the latter reflected Japan's need to ensure that it was not placed on the Arab oil exporters' embargo list. Japan's status *vis-à-vis* the international oil industry at that time and the problems associated with that status resembled in kind those of Western Europe (see Chapter 5), but in degree they were significantly accentuated by the interplay of a set of factors specific to Japan. These will be examined in this chapter.

Japan's exposure to the influence of the international oil industry was a relatively recent one. Its pre-war economic development and its wartime exertions were, like those of Western Europe, essentially coal-based. Up to 30 million tons of coal per year were produced in Japan and met the greater part of the country's energy needs. It was supplemented by two other indigenous sources of energy: first, by a great deal of 'non-commercial' energy, which may be defined as any combustible materials which could be collected or gathered (for example, wood or straw) by the peasants to provide them not only with their limited requirements of heat and light but also with 'energy crops' which could be sold to urban dwellers; and second, by hydro-electricity, particularly from the 1930s, when a great deal of the country's lowest-cost hydro-electricity potential was developed. In addition, coal was imported to sustain Japan's increasing industrial energy use in the years before 1941 from the rich coal-producing areas of Korea and China which Japan had conquered, partly with this objective in mind. Thus, the country's needs for oil up to and into the wartime period were limited mainly to the relatively small amounts required for motor transport and for the armed forces. During the war the country's small indigenous oil resources, plus oil from the fields which were captured in the Dutch East Indies and Burma and whose installations were repaired sufficiently during the Japanese occupation to yield some oil, met the needs of the fighting forces. In this respect Japan had fewer problems over essential oil supplies during the war than Germany, which was obliged to depend partly on expensive oil from coal technology.

In the immediate post-war period the initial rehabilitation and growth of Japan's industry and transport had to be based mainly on the energy that was indigenously available – coal and hydro-electricity. The latter was quickly brought back into full production, and even expanded, with the help of the U.S.A. The former depended partly upon mines that were approaching the end of their useful life and on an ageing mining force which was relatively unproductive. In these respects the Japanese coal industry's problems were similar to those of post-war Britain and Germany. In addition, the mines of Hokkaido and Kyushu, Japan's principal coal-producing areas, were remote from the main industrial areas (see Map 8, p. 147). Nevertheless, in spite of all the difficulties, the coal had to be secured in order to meet the country's minimum energy needs and government help was given to reactivate the privately-owned mines. Production increased, but by no means fast enough to satisfy the demands of an economy which, by the end of the 1940s and the beginning of the 1950s, was beginning to show great potential, partly as a result of the enforced changes which were made in the structure of Japanese society during the American occupation. Somewhat reluctantly, therefore, but faced with the need for additional energy supplies, and with the encouragement of the U.S. military administration used to oil as the main source of energy, the Japanese started to encourage the expansion of the oil industry.

In order to encourage the consumption of oil, financial assistance was given to power stations and to other large industrial users to convert to oil, whilst the problem of making the required products available at as low a cost in foreign exchange as possible was first tackled by the rehabilitation and massive extension of the old-fashioned, small and inefficient oil refineries. As in the case of Western Europe, crude oil could be imported at a lower foreign exchange cost than oil products, and in the face of the country's crippling shortage of foreign exchange – and, temporarily, of the means of earning it – crude oil importation certainly had to be encouraged. But even the expansion of these refineries caused economic problems because of the amount of capital needed, and the fact that much of the investment had to be spent on equipment and

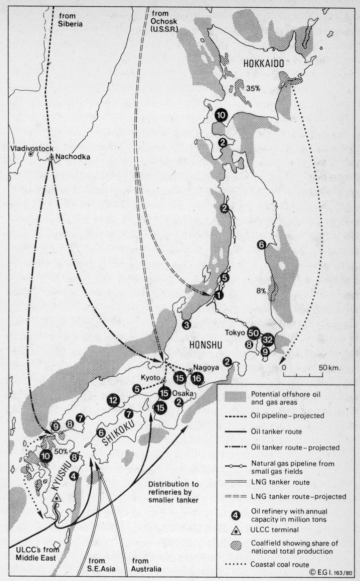

from Siberia

from Ochosk (U.S.S.R.)

HOKKAIDO

35%

⑩

②

Vladivostock

Nachodka

②

⑥

⑤

①

③

8%

HONSHU

Tokyo ㊿ ㉜

⑧ ⑨

Nagoya ②

Kyoto ⑮ ⑯

⑤ ⑮ Osaka

⑫ ⑮ ②

⑦ ⑦

⑨ ⑧ ⑥ SHIKOKU

⑩ 50% ⑧

④ KYUSHU

0   50 km.

Distribution to refineries by smaller tanker

▨	Potential offshore oil and gas areas
– – – –	Oil pipeline – projected
———	Oil tanker route
–·–·–	Oil tanker route – projected
—○—	Natural gas pipeline from small gas fields
═══	LNG tanker route
≡≡≡	LNG tanker route – projected
④	Oil refinery with annual capacity in million tons
△	ULCC terminal
▨	Coalfield showing share of national total production
·······	Coastal coal route

ULCC's from Middle East

from S.E.Asia

from Australia

© E.G I. 163/80

Map 8. The infrastructure of Japan's energy supply in the mid-1970s at the time of peak energy demand

expertise which at that time were available only from the U.SA. Thus, problems of capital availability and foreign exchange scarcity seemed likely to be a stumbling block to this solution to Japan's energy 'crisis' of the early post Second World War period.

In 1950, therefore, the government took a decision which it later came to regret, but which was unavoidable at that time. Contrary to the traditional Japanese attitude towards foreign participation in the country's basic activities, the government agreed, as a result of American pressures and the importance of the energy problem, to allow the major international oil companies to participate in the Japanese refining industry by linking up with the local refining companies. Each party put up half the capital required for the expansion of a specific refinery, with the contribution of the oil companies covering the foreign exchange costs of the investment. The international oil companies themselves had little enthusiasm for this procedure, to which they also only agreed somewhat reluctantly, under pressure from the U.S. military authorities, as they would have preferred to work through wholly-owned refining subsidiaries in Japan, as they did elsewhere, most notably in Western Europe, where refineries were also being expanded very rapidly. In return for their agreement to the joint ventures with Japanese companies the international oil corporations which were putting up the money and the necessary expertise secured the permanent right to supply all the crude oil needed by the refineries. We shall return to examine the subsequent criticism in Japan of these arrangements, but, in the circumstances of the time, they probably represented the only viable alternative to allowing the establishment of wholly foreign-owned refining subsidiaries as a means of achieving the rapid and efficient expansion of the Japanese oil-refining industry. In the event the expansion which was achieved enabled Japan to rid itself quickly of dependence on increasingly expensive and unreliable domestic coal. Within the space of a very few years the country was converted from a coal-based to a largely oil-based economy, and from a high-energy-cost to a low-energy-cost economy, with all the advantages that this brought to Japan in its competition for overseas markets for manufactured goods after 1960.

In 1950 oil contributed only 7 per cent of the nation's total energy requirement, while coal contributed over 60 per cent (the balance being mainly hydro-electricity). Only a decade later, in 1960, oil was just about as important as coal in the much expanded economy and its share of the total was, moreover, still growing rapidly. Indeed, as early as 1955 the government had had to reverse its policy of giving assistance to firms which changed to oil because the markets for coal were disappearing too quickly. In spite of the relatively high prices at which oil was transferred by the international companies to their associated refining companies in Japan at this time, based on the same pricing arrangements for international oil trade as described in the case of the Western European situation in the previous chapter, oil products were still able to undercut the use of coal in most energy markets. Thus, in 1955 statutory restrictions were placed on the installation of fuel-oil boilers to replace coal-fired boilers. These restrictions applied even to thermal power stations, which by then were desperately seeking to get access as soon as possible to lower-cost oil fuel to replace the coal on which they had hitherto depended. The trend to oil could not, however, be reversed by such steps, for the price differential in its favour over coal continued to widen. On the one hand, coal prices continued to rise under the stimulus of rapidly increasing wages paid to miners (labour costs are the most important element in mining in most parts of the world, and even more so in Japan, where labour productivity in coal mining was particularly low), and because of the increasingly difficult physical conditions for mining coal. On the other hand, oil prices in Japan were falling as a result of the competition between the growing number of importing and refining companies seeking markets, and the availability after 1957 of crude oil on the world market at prices below those formally posted by the international 'majors'.

Even in the earlier part of the 1950s there had been high annual rates of increase in oil consumption (when they averaged about 15 per cent per annum), but these growth rates were overshadowed by the 'explosion' in the rate of increase in oil use after 1957. Between 1957 and 1960 consumption doubled, and right through to 1964 the oil growth rate averaged almost 25 per cent per year. Even then, however, the period of rapid growth in oil demand, and its increasing relative importance in the total energy requirements of Japan, was by no means over. The continued rapid growth of the economy and the continued, even the increasing, ability of oil to compete with the coal that was still being produced and with the remaining, less attractive opportunities for the development of hydro-electricity potential still produced annual rates of increase in oil demand of more than 10 per cent. Indeed, rates of growth throughout the period remained so consistently high that all forecasts of the future demand made in the 1950s and the 1960s proved to be too conservative. A 1962 estimate, for example, indicated what was at the time thought to be an 'impossibly' high demand for almost 100 million tons of oil by 1967. Actual use in 1967 turned out to be about 108 million tons. Thereafter, until 1973 growth rates continued to be higher than those written into official estimates and in 1974 Japanese oil consumption broke through the 200 million ton level. In 1957 consumption of oil had been less than 15 million tons. At that time Japan was a modest seventh amongst the world's nations in terms of oil use: by 1974 it was third, lying behind only the two super-powers, the United States and the U.S.S.R., each with a population some two and a half times that of Japan.

In the light of the growth in oil use which Japan experienced, the country's newly created Energy Commission attempted in 1967 to make forecasts for the growth in energy and oil demand up to the year 1985. There was, of course, at that time a general expectation that economic growth would continue unabated under the stimulus of the expansionist forces which in 1967 were so clearly at work, and which provided the kind of development which both Japan and the rest of the world appeared to want. In light of this the Commission forecast a total demand for energy by 1985 equivalent to over 800 million tons of oil (compared with demand at the time of the forecast of about 200 million tons). Moreover, it also concluded that the share of oil in the total would rise from under 60 per cent to over 75 per cent so giving a 1985 oil requirement of over 600 million tons. Even this estimate was thought by some to be too pessimistic, for the Energy Commission assumed that nuclear power would grow to meet almost two-fifths of the balance of the energy needed. Given the considerable difficulties that still existed in building successful nuclear power stations – particularly in a country as earthquake-prone as Japan – this view of nuclear power was generally felt to be unachievable, and as oil was the only possible alternative this implied that oil demand in 1985 could be even higher than the Energy Commission's 600 million ton forecast.

As this brief survey shows, the Japanese economy quickly became one whose well-being was inextricably tied to the oil industry. One should note, however, that just a few years after the Energy Commission's 1967 forecast, Japan started to worry a great deal about the environmental impact, in respect of both atmospheric and water pollution, of such an intensive use of oil. Between 1970 and 1973 measures were taken to limit these adverse effects, particularly as far as requirements for air-pollution controls were concerned, and there were already indications by 1973 that these were beginning to slow down the rate of increase in oil use. Since then, however, this motivation to restrict the use of oil has been greatly strengthened by the oil supply crises arising out of restraints on production in countries of the Middle East (on which Japan mainly depends for its imports), and by the rapid rise in oil prices since October

1973 (see Figure 1, p. 142). The impact of these factors was to heighten the degree of concern for efficiency in the use of oil and to encourage the use of substitutes. As a result, rates of growth in oil use fell quite dramatically by the late 1970s, to an average of about 3 per cent per annum (an additional factor in this respect was, of course, the much slower rate of economic growth in the post-1973 period of economic difficulties in the world system). Since 1979, however, the use of oil has declined very sharply, in spite of continued economic growth at an average 3 per cent per annum rate. As a result Japan used less oil in 1985 than it used in 1974 (under 200 million tons) rather than the minimum use of 600 million tons of oil per year by 1985 as forecast in 1967. Even more recent forecasts failed to take the propensity of Japan to use less oil per unit of production sufficiently into account and still suggested a 1985 oil consumption of about 250 million tons.

For Japan, however, it remains highly unlikely that there can be an increasing contribution to the country's energy economy by its indigenous coal industry, as was expected in the case of the U.K. and Germany. Japanese planners hope that coal production can be maintained at about 50 million tons per year and this should be compared with the serious doubts before the 1973 oil crisis as to whether even 20 million tons of Japanese-produced coal could really hope to compete in the country's large energy market. The British coal industry reduced to the same proportionate share of the energy market would have been no more than a quarter of its size in the early 1970s. Even the most pessimistic estimates for U.K. coal output did not visualize it being reduced to less than 50 per cent of its 1971 level of 150 million tons. Even more significant, Japan did not have the immediate promise of the very large resources of natural gas on which Europe could count, as shown in the previous chapter, and the increasing use of which has served to undermine the growth prospects for the European oil industry.

Western Europe has also been able to reduce its oil imports by the exploitation of indigenous oil resources, but no similar large-scale oil resources have yet been found in Japanese offshore waters. Finally, Japan did not have an existing nuclear power industry, with many years of experience in producing nuclear electricity and with the domestic technology and know-how that make nuclear power virtually an indigenous fuel, as in the case of the countries of north-west Europe. Japan will, of course, achieve this, but in the meantime nuclear development has, of necessity, been relatively slow, particularly given the need to build nuclear reactors that will withstand the earthquakes to which much of Japan is prone. Therefore, in spite of the fundamental changes in the international oil supply and price position, Japan's continued relatively high level of reliance on imported oil has continued. This situation justifies the conclusion, mentioned at the beginning of this chapter, that no country outside some of the major oil-producing nations has had more to be concerned about in respect of developments in the world of oil power since the early 1970s.

It is not surprising, therefore, that successive Japanese governments after 1950 worried about oil to an increasing degree. Their degree of concern was, moreover, gradually reflected in an increasing willingness and intention to do something about it. In the early post-war period official concern lay principally with the foreign exchange cost of oil imports and thus the industry was seen to merit the special treatment that it received – that is, official blessing for joint ventures, whereby foreign capital was, in part, used to finance the expansion in oil refining. Apart from this special treatment, oil was also an important element in the general system of control exercised by the government over all imports. Because of Japan's acute shortage of foreign exchange a complex government system of allocating funds for imports was gradually developed and, as part of this system, the oil industry had to argue its case for foreign funds twice a year and have its arguments weighed against the claims of other sectors of the economy. Because oil was essential to the country's recovery and because its foreign exchange cost was not a dominant element in the total foreign exchange requirement, the industry was usually given high priority and so met only temporary difficulties in the occasional periods of financial stringency through which Japan passed in the fifteen years after the end of the war. Nevertheless, from accounting for only about 8 per cent of the value of imports in 1950, oil's share of the total rose gradually through the decade until it reached 16 per cent in 1960. This level was reached, moreover, in spite of the falling real crude oil prices that Japan enjoyed in the second half of this period.

Thus, the government used its foreign exchange allocative powers to ensure that preference was given to refinery proposals which incorporated plans for, or the possibilities of, petrochemical developments. The encouragement of this industry was, of course, another means of improving the country's balance of payments, both by important substitution and by the development of an industry with export potential. From very small beginnings in the mid-1950s, petrochemicals in Japan were thus encouraged to expand. Expansion took place so quickly that by 1966 the Japanese petrochemical industry was second only in size to that of the U.S.A. in production and in exports. From four small complexes in 1958 the industry expanded to eleven major ones by 1978, all associated with the oil-refining industry of the country. Thus, oil imports were made to pay for themselves in terms of their foreign exchange costs. Crude oil imports were in essence viewed as the raw material to which much value could be added in manufacturing in Japan. The resulting output of large quantities of petrochemicals could, in part, be exported and so secure export earnings in excess of the foreign exchange costs of all crude oil imports.

The method by which foreign exchange for oil refiners was allocated was designed to encourage the companies to search the world's producing areas for the cheapest supplies of crude oil and, similarly, the methods of apportioning foreign exchange allocated for the import of refined oil products (principally fuel oil, which the Japanese refineries did not produce in sufficient quantities to meet the heavy peak demands) also sought to stimulate the purchase of low-cost supplies. The ending of general import controls in 1962 when, by international agreement, Japan agreed to liberalize its trading régime, foreshadowed an end to this system of close government control, not only over the oil industry's imports in particular, but also over its activities in general. This was viewed with official consternation in the light of the increasing significance of oil in the Japanese economy and the knowledge that these essential supplies of energy were really out of the control of the Japanese themselves, and in the hands of a small group of international – mainly American – oil companies and an even smaller group of oil-producing nations in a politically unstable part of the world.

About 80 per cent of Japanese oil imports were tied – by permanent contracts – to those few international oil companies whose financial assistance in promoting refinery expansion the government had been obliged or persuaded to accept in the earlier post-war period. Apart from the initial need in 1950 to accept financial and technical help from the United States to get its oil-refining industry off the ground, the continued rapid growth in the demand for oil products in Japan throughout the 1950s and into the 1960s put such a severe strain on the local capital market for financing the infrastructure expansion of the industry that all Japanese refining and marketing companies were more or less continually obliged to accept loans from the international companies to finance development or expansion. In return the international companies involved sought and obtained additional guaranteed outlets in Japan for their increasing availabilities of crude oil from the main producing countries. To make the degree and dangers of dependence on external institutions even worse, over 80 per cent of Japan's essential oil imports derived from the oil-producing countries of the Middle East. Japan thus found itself dependent upon the goodwill of the governments of Kuwait, Iraq, Saudi Arabia and Iran for the security of its oil supply.

In view of the obvious economic, political and strategic implications which were being generated by the rising tide of oil imports, the Japanese government decided that it could not afford to 'liberalize' oil. It resolved, therefore, to replace the control it had had over oil through the import system with a direct control of the industry exercised through new oil legislation. Thus, in 1962, concurrently with the Japanese liberalization of its trading régime, a new law was passed which specifically related to the oil industry. Though this was essentially 'enabling' legislation, it did give the government the possibilities of very wide powers by means of which it could control the oil industry in almost all its activities. The degree to which these powers would be implemented was made dependent on the outcome of the more or less continuing discussions between the government and its agencies on the one hand, and the different sectors of the oil industry on the other.

The new law not only provided for the continuation of the measures of control over prices and refinery expansion that had previously existed, but also enabled the government to insist on outlets being found for all home-produced oil and any oil which was produced by Japanese companies operating overseas. Its intention was, first, to limit the growth in the foreign exchange cost of the commodity and second, to curb the power exercised by the foreign companies and governments over Japan's energy. But this was not all, for the law, as passed by the Diet, which had sought to strengthen it beyond the intentions of the government, added further scope for building up Japan's position in the oil world. Thus, the law created the possibilities of financial aid and tax exemption for the exploration and development of Japan's own oil and gas reserves; it promised assistance to Japanese refiners who sought more freedom in making their crude oil purchases by terminating or modifying the crude oil supply agreements to which they had committed themselves with the international oil companies; and it allowed the establishment of a national corporation to market indigenous and Japanese-produced overseas crude oil – termed 'classified crude' – developed by national companies. Powers were also given to this corporation to direct such classified supplies to other refiners.

The readiness, willingness, ability or need of the government to implement in its entirety such a far-reaching piece of oil legislation did not, however, materialize immediately. In fact, reporting in 1966, the Petroleum Committee of the Energy Council had to note that little had been achieved in fulfilling government policy, which in theory sought to promote low-cost secure supplies which would be provided, at least in part, by fully integrated Japanese companies. In the years 1962–5, the new legislation made little difference. Thus, crude oil import prices, which had fallen year by year from 1957 to 1962, thereafter remained almost constant, in spite of a growing availability of crude oil in the world. 'Tied' crude still accounted for 80 per cent of total supplies and was permitted in even at an average cost per barrel which was 10 cents (about 5 per cent) higher than 'free' crude oil. There had been little development of overseas ventures, apart from the gradual development of the field discovered by a Japanese consortium in the Neutral Zone in the Persian Gulf in 1959. Initial development of this field – the Khafji field – had been rapid by world standards, but then output was restricted by the unwillingness of refineries in Japan controlled by foreign companies to take more than a very limited amount of its crude oil. The company – Arabian Oil – almost in desperation because of its need to achieve a higher cash flow from its investments, tried to sell an interest in the field to American companies, in return for their being given the right to take up to 20 per cent of the crude oil available. Thus, oil production, developed initially to feed the Japanese markets with Japanese-produced oil, was being forced into other markets because no room could be found for it in Japan – even with Japanese demand expanding at a world record rate of about 20 per cent per annum! By 1966, Khafji crude still provided under 10 per cent of Japan's total crude-oil supply. Finally, attempts to use Soviet oil as an alternative to the predominant Middle East international oil companies' supplies had been little more than half-hearted, and in 1966 it still provided under 5 per cent of the total.

All this the Petroleum Committee of the Energy Council found depressing, expensive for Japan, and not in the country's best political and strategic interests. It thus recommended several lines of action, all of which involved much greater government intervention in the oil industry. First, it called for a 10 per cent reduction in the 1966 level of crude oil import prices by 1970. Second, it recommended a reduction in the proportion of 'tied' oil imported by Japanese refineries to 50 per cent or less of their requirements, justifying this figure by reference to the maximum equity interest of 50 per cent that the supplying companies had in these refineries. Third, it also suggested that the state should be responsible for overseas oil exploration and development ventures and for promoting indigenous sources of supply so as to bring their contributions up to 30 per cent of the total oil requirement by 1985. And, fourth, it advocated larger imports of Soviet oil. These, it thought, could be achieved through a pipeline from the Siberian fields to the Pacific coast of the U.S.S.R., with the U.S.S.R. supplying the oil in return for deliveries of Japanese steel and other equipment for the line (see Map 8, p. 147).

The government decided to take action on these recommended directions, even though it knew that, in doing so, it risked conflict with the international companies. Japan felt strong enough, however, by 1967 to face such risks, especially as it had seen how many European countries had gained what they wanted from the same oil companies through a policy of toughness rather than of appeasement. Refining companies were thus gradually 'persuaded' to accept more Khafji crude. Negotiations on the import price levels of other crudes became more or less continuous and prices were gradually brought down, until the beginning of the period of constrained international supply in the early 1970s. Negotiations with the Soviet Union over supplies from Siberia, via a pipeline to the Pacific coast, were re-opened and were given more impetus by the disruption of Soviet supplies from the Black Sea to Japan caused by the closing of the Suez Canal after the Middle East war of June 1967. The negotiations were successfully concluded with an agreement in principle to go ahead with the project. The effect of this agreement, had it become fully operational, would have been to increase Soviet oil supplies to about 15 per cent of Japan's total imports. The project was, however, delayed for a variety of reasons so that Japan has not in fact taken such large volumes of Siberian oil. Meanwhile, the success of a joint Soviet/Japanese oil exploration effort off the shore of the island of Sakhalin (to the north of Hokkaido – see Map 8, p. 147) appeared to offer an alternative source of supply for Russian oil to Japan, though this too has so far failed to materialize as it has had a lower priority in Soviet oil-development plans than exploitation elsewhere in the country. In the meantime, the problem of transporting Soviet oil to Japan, which arose from the closure of the Suez Canal in 1967, was resolved in a way which clearly illustrated the 'internationality' of the oil industry. Iraq and B.P. joined Japan and the U.S.S.R. in solving it. Soviet oil to Japan was replaced by supplies of B.P. oil from Iraq, whilst B.P., in turn, took an equivalent amount of Soviet crude oil out of the Black Sea ports for use in its markets in Western Europe, whose normal supply of crude oil from countries East of Suez had also been disrupted by the Canal closure!

Of greater importance, however, in revising and strengthening Japanese oil policy after 1966, was the series of decisions taken to try to increase the supplies of oil under Japanese control, and hence available at a lower foreign exchange cost. In October 1967 a Petroleum Development Public Corporation was formally inaugurated to coordinate and promote development and production by Japanese companies. It was authorized to give direct financial aid and to guarantee loans from other sources, to lease exploration equipment, and to give whatever technical assistance and guidance might be required. The cost of the programme, designed to lift the Japanese-owned share of Japanese oil imports to 30 per cent of a steadily increasing total, was evaluated, in terms of cost levels in 1967, at £1,500 million by 1985, and this, together with a shorter-term plan for exploration and development overseas to cost over £400 million by 1975, was to be financed under arrangements made by the Ministry of Industry and Trade, which hoped to combine private funds with government loans and other assistance.

There was little doubt by this time that the Japanese government really meant business in this direction and the steps announced had by 1970 already produced significant results. Khafji crude oil was entirely earmarked for Japan and attempts by the company to sell it elsewhere were not approved. In addition, several other positive steps were taken to generate Japanese owned or controlled oil production. For example, an interest was bought in a Canadian company for a joint exploration programme in Canada; a link was formed with an American company to investigate the potential for developing the Canadian Athabasca tar sands; an Alaskan Oil Resources Development company was formed, and it joined forces with Gulf Oil to exploit acreage in Alaska; the Indonesian Petroleum Development company was expanded and both government and private funds were found to finance the development of promising off-shore discoveries in north Sumatra and Kalimantan; an interest was bought in an Australian company for work both in New Guinea and Australia itself; and, finally, various exploratory negotiations were started in several Latin American countries.

Thus, Japan's foreign oil ventures by 1973, even before the oil supply and pricing crisis of that year, under the auspices of the government-financed Japan Petroleum Development Corporation, already totalled more than fifty and added up to very much more than the isolated and somewhat perfunctory efforts of the earlier period up to 1966, when activities were still limited to the Persian Gulf Neutral Zone and to an agreement with Pertamina, the Indonesian state oil entity.

The Japanese, indeed, committed themselves to large exploration expenditures in many parts of the world – though most notably, in the first instance, to developments in Abu Dhabi in respect of two separate projects. The first involved expensive competitive bids for on-shore and off-shore acreage in the promising oil-bearing zones of Abu Dhabi in the lower Persian (Arabian) Gulf. These concessions were won by the consortia of Japanese companies, in the face of international competition from established oil producers of the U.S.A. and Western Europe, and the successful Japanese bids involved total minimum exploration expenditures of almost $50 million by 1975, even if no oil was discovered at all. Given the nature of the concession areas this seemed highly unlikely, for they were essentially extensions of existing producing areas of high potential. Indeed, the Japanese companies struck oil almost immediately and then developed the finds quickly, so that oil from Abu Dhabi produced by the Japanese consortium could make a contribution to the needs of Japan's refineries by the middle of the 1970s. This flow of oil from Abu Dhabi to Japan was then further increased as a result of the second Japanese project there. This involved the purchase of a 50 per cent share in B.P.'s proven oil rich concession in Abu Dhabi at a cost of almost $600 million. The price paid appears to have been related to the likely ultimate availability of reserves in the concession amounting to no less than 80 milliard ($= 10^9$) barrels of oil, the development of which would have been of no small importance, even in relation to Japan's large and rapidly growing requirements for crude oil imports. The profits from their development, moreover, were expected to enable further exploration and development work to be self-financing.

The late 1970s and the early 1980s could well thus have been marked by as rapid a growth in Japanese financed and controlled oil production as the previous ten years had seen in Japanese oil consumption. The likelihood that this would happen was made stronger if one recalls the size of the oil market which was anticipated by the mid-1980s. At the time of this move by Japan into large-scale foreign oil operations, this was expected, as we have shown, to be at least 600 million tons per year. This Japanese aim has been thwarted not, as might have been expected given the very strong reaction in Japan against the foreign oil companies, by increasing difficulties in Japan's relationships with the international oil companies, but rather by the actions of the oil-exporting countries in nationalizing their oil industries, including, of course, the Japanese companies involved. Thus, in respect of the traditional oil-producing and oil-exporting nations of the world Japan was thwarted in its efforts to secure control over its oil supplies in the way in which it planned. As with most companies which have had their oil-producing operations nationalized, however, the Japanese corporations concerned have been able to sign technical, managerial and oil-supply agreements with the oil-exporting countries concerned and these, of course, have provided some element of security for Japan in terms of its oil needs, particularly in the context of import volumes, which are now lower than they were in 1979.

Apart from this, however, the post-1973 changes in the world oil outlook have necessitated a reappraisal by Japan of the options open to it. First, of course, as already indicated, it has cut back significantly on its oil use – partly by improving the efficiency of its use and partly by diversifying into other energy sources. Such deliberate diversification has included the import of large volumes of coal from what are considered to be politically safe countries, especially Australia, South Africa and Canada, and the use of imported liquefied natural gas from Indonesia and other countries anxious to build up their sale of this new source of energy. Second, it has pursued very active diplomacy in efforts to improve and stabilize its relationships with oil-producing countries, particularly those of the Middle East. Specifically in this respect it has agreed to develop, on a significant scale, direct state-to-state trade in oil. The national oil companies of several of the producing countries have sales, transport and even refining agreements with Japanese oil corporations. Third, it has extended its search for involvement in the development of oil production capacity to many non-O.P.E.C. countries, usually in the context of production-sharing agreements which do not necessitate concession-type arrangements. Fourth, it has concluded negotiations with China for an expanded trade agreement involving the exchange of Chinese oil for Japanese capital and consumer goods. Given China's undoubted potential as an oil producer this could be a most significant development and the prospects of China ultimately supplying Japan with up to 100 million tons of oil per year cannot be discounted, though this can not happen before the 1990s, given the continuing difficulties in China in increasing oil output rapidly in the shorter term. And fifth, home-based production of oil – and natural gas – is being given serious attention in spite of what appear to be limited geological prospects.

There was a successful discovery of natural gas in the Sea of Japan a few years ago – as a result of a joint venture between a Japanese company and a Shell subsidiary (this being the first time that a non-Japanese company had been permitted to participate in a mining and oil exploration venture in Japan itself). Other companies have now joined in the exploration of Japan's close continental shelf with a number of other successes, but so far there has been no indication of a potential for the discovery of large oil and gas resources. But the search continues and is even being intensified. Apart from this, however, more distant parts of the east Asian offshore regions in which the Japanese claim an interest have been brought into the reckoning. This involves agreements with the neighbouring states of the U.S.S.R., North and South Korea and China. Such agreements are not in sight except in the case of a median line agreement with South Korea, where, as a result, initial exploration work has already been carried out and a number of wells have been drilled – so far without any success. In the meantime, the older joint-venture agreement with the Soviet Union in respect of petroliferous areas off the island (part Japanese and part Russian) of Sakhalin has produced a number of major discoveries and this may prove to be the first significant possibility for developing indigenous (or partly indigenous) oil resources for Japan in spite of Soviet reluctance to commit large investment to this development (for the reasons indicated above) in the short term, and because of problems with ice conditions in the sea area involved.

In all these developments the role of the traditional oil companies in supplying oil to Japan has become much diminished and their position is becoming more uncertain. While the Japanese market for oil continued to grow the possibility of a serious clash with the international majors over the new oil policy was unlikely. There was room for everyone in a growing operation, and profits for all that were efficient. In the context of a stagnant or a declining oil market in Japan (as since 1979), a potentially serious conflict of interest between the state and the international companies is clearly possible, in much the same way as we have already seen it at work in Western Europe, as well as in the producing countries. The pathway that the companies have to tread to avoid this conflict is a narrow and dangerous one, and one false step in Japan – as in many other countries – could see them effectively squeezed between opposing nationalisms. In the case of Japan this would be particularly harmful to the companies, for here is the world's largest and most secure market for international oil for as far ahead as it is possible to judge, in spite of the shock administered to the Japanese economy by the traumatic events in the international oil world since 1973. In the light of this, one may expect that the companies will make every effort to continue to be involved in supplying oil to Japan and in refining and marketing it there. They have, in addition, sought and achieved involvement in new opportunities, such as the possibilities of developing indigenous oil and gas production, and in new endeavours, such as the supply of coal and of liquefied natural gas to the country. Increasing imports of such alternative sources of energy, particularly from other parts of the Far East and from Australia, are substituting the use of oil so that the international oil companies' continuing involvement in the trade is a prerequisite to their continuing success in the Japanese energy sector in the coming decades.

7. Dependence on Oil in the Developing World

This chapter is concerned with the remaining group of countries in the world of the international oil industry. This group consists of those countries which have interests in common arising from their under-development and low living standards and which are dependent nations in the economic system of the non-communist world. Often an under-developed country's participation in this system is only marginal, con-sisting of little more than the export of one primary commodity or another to the industrialized countries in return for which foreign manufactured goods are imported for use by the minority of the popu-lation of the country able to buy such things. At the same time many, or even most, of the people of such countries pursue their own subsistence or largely subsistence ways of life and, in the light of their very rudi-mentary wants, are virtually unaffected by such external trading rela-tionships. There are a very large number of such countries – well over 100 in Africa, Latin America and South East Asia – and their general economic and social problems have been investigated in much post-war literature. Our particular concern with them in this book is as consumers and potentially much larger consumers of oil, and in this respect one has first to note that almost all of them use oil as the most important source of energy in the modern sectors of their economies. It should be borne in mind, however, that in most such countries the 'energy' that is collected or gathered within the framework of a subsistence way of life remains more – and often much more – important than all purchases of all types of 'commercial' energy put together. The dominance of oil in their commercial energy systems is illustrated in Map 9 (p. 162). The dominance of oil so early in the development process differentiates them from the industrialized nations where, as we have seen in earlier chapters, coal was in most cases the main source of energy which sustained econ-omic growth. The switch away from coal to oil occurred in quite recent years and, since 1973, the process has, in part, been reversed. A switch to coal is not an option which is open to most of the developing countries in the short-term so that their economies have been made more pre-carious than those of any other group of countries in the world as

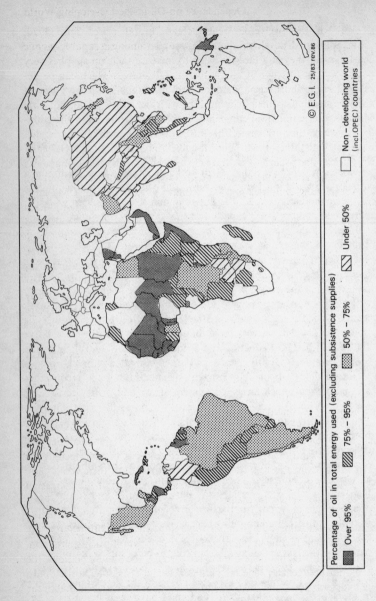

Percentage of oil in total energy used (excluding subsistence supplies)

▓ Over 95% ▓ 75% – 95% ▓ 50% – 75% ▨ Under 50% ☐ Non – developing world (incl. OPEC) countries

© E.G.I. 25/83 rev.86

Map 9. Dependence on oil in the developing world.

a result of the twelvefold real increase in oil prices since 1970.

Since almost all developing countries are also aiming at rapid economic development, they still face the need to increase their oil supplies very quickly, in the absence of alternative energy sources which can be developed in the short term. Many of them have thus had to give a great deal of attention to their relationships with the world of oil, not only with the international oil companies and the governments of the countries in which they are based (the U.S.A., Britain and the Netherlands), but also with other large countries like Italy, France and Germany which can offer some lessons and help in dealing with the oil companies and with the oil-producing nations. Incidentally, the latter nations are also developing countries and, as we have already seen in Chapter 4, they have had problems specially related to their function as producers of a single primary commodity. They do have, however, unlimited quantities of oil and gas available for their own use, a highly favourable balance of trade position as a result of their oil exports, and large government revenues arising from the payment of oil royalties and taxes by the oil-producing companies. And, as we have previously seen, their ability to charge very much higher prices for oil since 1973 has largely eliminated the problems that they might earlier have been said to share to some degree with those countries of the Third World which do not export oil.

The developing countries we are concerned with in this chapter lack the considerable advantages of the oil producers and, as will be shown later, they increasingly consider themselves the means, in part at least, whereby the oil-producing countries have secured their wealth. To the governments of many underdeveloped oil-consuming nations it long appeared that the international oil-system was so arranged as to transfer income from themselves to the oil companies and the wealthier oil-producing countries. Their beliefs have been more than confirmed by the post-1973 adverse effects of the higher oil prices on their economies. They would like to change this situation, in the light of the increasing gap in incomes *per capita* that has opened up between the developing oil-producing nations on the one hand and the developing oil-consuming nations on the other but, given an apparent inability to bring about any change in their favour in the present international oil-pricing situation, in spite of efforts that have been made through the United Nations, the development agencies' special conferences such as the North–South dialogue in 1976–7, and through special investigations such as that of the Brandt Commission in 1981, the discrepancies are still getting larger rather than smaller.

The developing oil-consuming nations do share interests in common – over the oil price question, for example – with the consuming nations of Western Europe and Japan (see Chapters 5 and 6). These latter nations are, however, relatively much more powerful, both from the point of view of their political status and also from their position as considerable consumers of energy. They were thus gradually able until 1970 to secure their increasing demands for oil at prices which moved downwards in response to their political and economic pressures (see Figure 1, p. 142). In contrast, the world's developing countries lacked, and still lack, both political and economic power. Politically, many of them are only recently independent and they are feeling their way with difficulty in a complex world in which the organization of the oil industry is but one of many problems with which their limited number of experienced diplomats and technical experts have to deal. Economically, they are such relatively small users of oil products that their markets are not necessarily attractive to competing suppliers, particularly when compared with the much bigger markets of Europe. Thus, their concern over oil and oil prices has been fraught with difficulties throughout the post-war period and their position with respect to the oil sector can in large part still be interpreted as a neo-colonialist one. This will be explained later in the chapter.

Before going on to that, however, we must examine in a little more detail the position of these countries as energy consumers, noting first that underdevelopment and a low use of energy go hand in hand. The general validity of this correlation has been established in numerous studies over the last thirty years or so. Thus, one finds that the *per capita* use of energy in countries like India, Brazil, Ghana and many others may be calculated in terms of tens or, at most, hundreds of kilograms of coal (or coal equivalent) per year, while more developed nations annually consume tons of energy per head. At the two extremes lie the U.S.A., with an annual *per capita* consumption of about 10,000 kilos of coal equivalent, and at the other, Nepal, using only about 10 kilos per head per year. Though one cannot talk of low energy use as causing economic underdevelopment, there is nevertheless some element of 'circular causation'. In this process bottlenecks in energy supply make economic growth difficult, while economic growth causes energy supply problems.

An example of the way in which such difficulties emerge and develop could be seen in the case of the Cauca Valley of Colombia in South America in the 1960s. In that region, under the stimulus of favourable conditions for both agricultural and industrial growth in the post-war period, the demand for energy far outstripped the ability of the local, small-scale coal industry and the public electricity authority to provide it. As a result industrialists in particular had to resort to costly expedients to secure sufficient energy in order to enable them to continue to operate. The cement company, for example, had to integrate back into coal mining to ensure its necessary supply of fuel, using some of its scarce managerial abilities and capital in an activity divorced from its main interests in manufacturing cement and cement products. Other manufacturing plants, unable to rely on the public electricity supply (which became overloaded at peak periods and thus had to cut off supplies to some customers) for their essential and continuous electricity needs, were forced into the expensive expedient of providing their own small diesel-generated units, the fuel costs of which *alone* were higher than the price the manufacturers had expected to pay for public electricity. With such bottlenecks, and the kinds of efforts and costs incurred by consumers to overcome them, it was only to be expected that potential development was thwarted by the unwillingness of new firms to put themselves in the same position, or by the fact that existing firms could not expand as quickly as they might otherwise have done. So the pace of economic development was slowed down.

Such energy bottlenecks occur particularly when developing countries make a start along the path towards economic development, for this inevitably leads to a period in which there is a rapid rise in the rate of energy consumption. As societies emerge from a largely immobile, subsistence-type economy into one in which transport, both of goods and people, becomes more significant, in which urbanization becomes of greater importance and in which industries are established both to substitute imports of manufactured goods and to make use of the country's resources, so the demand for energy starts to rise rapidly. Transport – particularly motor transport, which is nearly always given priority for investments these days – is very energy intensive. Urbanization, which involves taking men and their families from a rural peasantry or labour force, means that families which previously collected or produced their own very limited energy requirements for simple heating and lighting purposes, now crowd together in the slums of the emerging urban centres and have little access to wood, other vegetable matter or animal dung for their energy needs. Instead, they come to rely on kerosene or, at a somewhat later stage, when they are officially connected to, or can unofficially tap, the electricity distribution network, on electricity for heating, lighting and cooking. Then, as their mode of living changes from subsistence or semi-subsistence rural to fully urban, their demand for energy as a family rises far above the level to which they have been accustomed. Meanwhile, the emergence of industries creates a completely new set of demands for fuel and power. This is particularly the case in industries which are as energy intensive as iron and steel and cement. These are, however, just the kinds of industries that most developing countries establish as soon as possible, either to take advantage of the availability of local resources, or to reduce the cost of their imports of these products – or both.

It is for all these reasons that one finds in the post-war period – when numerous nations have embarked on such policies of development – that the most rapid rates of increase of energy consumption have generally been in Latin America, South East Asia and, more recently, Africa. In both Malaysia and Brazil, for example, the average annual rate of growth of energy consumption in the twenty years up to 1973 was of the order of 12 per cent. These rates are not untypical of those countries of Latin America and South East Asia where economic development plans, centred around industrialization, have been followed throughout the post-war period (or since independence, if later than 1945). In the case of most African nations, industrialization and development planning is a more recent phenomenon and the high energy growth rates date back only to the late 1950s or the 1960s. These very high growth rates in energy use reflect not only the growth of the economies concerned in terms of total output of goods and services, but also the increasingly energy-intensive character of their development. This often gave a consequent coefficient of energy use approaching 2·0 (this means that for each percentage point rise in economic activities there is a 2 per cent rise in energy use). This high coefficient of energy use in the developing countries may be compared with figures of only 0·7 and 0·85 for the U.S.A. and Western Europe respectively during the same period. In the latter countries the much lower coefficients reflected the lower needs for additional inputs of energy in the later stages of the development process. The way in which these relationships work out over time is shown in Figure 2 (p. 168).

One should, however, remember that the high percentage rates of growth in energy consumption for the developing nations also reflect their low levels of energy use at the beginning of the period. Even after fifteen to twenty years of rapid growth in energy use the absolute consumption levels that they have reached are still quite modest by American and Western European standards. Brazil's 120 million people still used only 65 million metric tons of coal equivalent (m.m.t.c.e) in 1980, compared with over 325 m.m.t.c.e. by the 50 million people of the United Kingdom. Malaysia, with a population in 1980 of about 13 million, used less than 10 m.m.t.c.e. in that year, while the Netherlands with a similar number of people consumed about 60 million tons. In total, in 1980 the world's developing nations outside the communist bloc, and also excluding the main petroleum-exporting countries, used approximately 1,100 m.m.t.c.e. – still less than half the amount of energy used in the U.S.A. in the same year.

The process of a rapid increase in the use of energy at the early stages of economic development is not, of course, a new one. The developing countries of the nineteenth century – Britain, Germany, France, the U.S.A., etc. – went through exactly the same process. For these countries, however, the development of their own coalfields provided the energy for economic growth, and the growing economic strength of particular nations and of particular regions was closely associated with the local availability of coal. Even as recently as the 1930s it was still generally considered that countries or areas wishing to follow in the economic footsteps of Europe and of North America would have to rely similarly upon the exploitation of indigenous coal resources. However, partly because of the dislocation in coal supplies during the Second World War and subsequent significant price increases, and partly as a result of the increasing ability and willingness of the oil companies to serve any markets in the world, demand for coal was replaced by a demand for oil. The idea that economic growth must rely upon the development of indigenous coal resources was tempered by the recognition that geological, technical and economic difficulties standing in the way of most attempts to expand coal resources quickly could be circumvented by using oil in almost all end-uses. The single important exception was the use of coke in blast furnaces, but even there technological developments enabled part of the coke charge to be replaced by other fuels. Even more significant in this respect are the steel-making processes which have been evolved to eliminate the traditional blast furnace entirely. It is not perhaps surprising that most of the progress made along these lines has been in Mexico, one of the fastest-growing of the developing countries, but one with very poor coal resources.

The increasingly widespread use of oil in the post-war developing world was assisted both by the inherent physical characteristics of oil and by the organization of the industry. Oil is easy and generally safe to handle, even with only limited technical resources and expertise, and it can be shipped around the world relatively cheaply in both small and large quantities. It could thus be made available wherever energy was required. In this respect it had, and, indeed still has, significant advantages over coal, particularly as far as ease of handling is concerned. At the same time, the relatively unsophisticated and divisible nature of oil transport and use

167

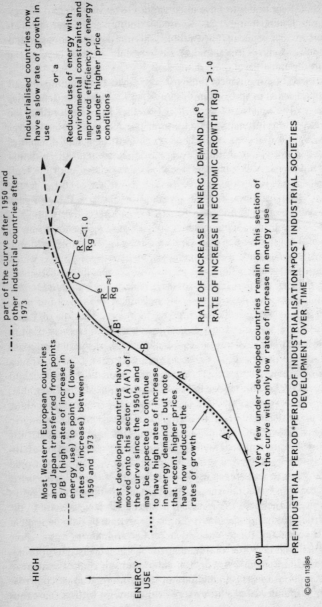

Figure 2. The relationship between energy use and economic development over time. This figure demonstrates how rates of increase in energy use in developing countries remain higher than the growth rates in the industrialized world. As developing countries depend mainly on oil for their energy this means that their importance as participants in the international oil industry will steadily increase.

ensures that it continues to have an advantage over more recent energy competitors, notably natural gas, whose distribution and use requires an infrastructure economically acceptable only in large-scale use, and nuclear power, whose technological requirements are still beyond the capabilities of most developing countries.

In addition, because the oil industry is organized internationally, all the facilities needed to make oil supplies available to developing countries could be easily provided. The ready availability of capital within the framework of the international oil companies meant that, no matter where the opportunity for investment arose, local financing difficulties did not stand in the way of the provision of facilities. The international oil companies were, moreover, equipped not only to acquire universal knowledge of investment opportunities, but also to act on them. In the case study to which reference has been made – that of the Cauca Valley of Colombia – the vacuum left in the supply of energy by the failure of the local coal industry to expand was quickly filled by the provision of oil by the local affiliates of some of the international companies – Mobil, Shell and Esso – in spite of the complicated supply arrangements which this necessitated. Oil products, in relatively small quantities, had to be lifted from the refineries on the Caribbean coast of Colombia (with their crude supply originating in the Middle Magdalena Valley little more than 300 kilometres – as the crow flies – from the Upper Cauca Valley). They then had to be shipped by coastal tanker via the Panama Canal to the Colombian Pacific port of Buenaventura, and then by rail – and later pipeline – over the Pacific Cordillera Range to depots in the Cauca Valley, from where the products could be distributed locally by truck to industrial and other consumers. Thus, in the Cauca Valley of Colombia, as in every other part of the world where they were permitted to operate, the international oil companies, with personnel used to and willing to work anywhere to further their company's business (and, of course, their own careers), sought and created opportunities for selling oil by establishing the terminals, depots, etc., needed for their operations. It would be very difficult to find a case anywhere in the developing world where, politics permitting, the international oil companies failed to respond to an energy 'gap' or failed to produce a profit out of their activities.

Thus, the developing world has gradually become heavily dependent on supplies of oil for its economic advance. In Latin America as a whole between 1950 and 1975, the total demand for energy increased by approximately four times, from less than 100 million tons of coal equivalent to over 350 million tons. Over this period the consumption of coal rose only from just under 10 to a little over 16 million tons, with its share thus falling from some 8 per cent of the total to under 5 per cent. The bulk of the increase in the overall demand for energy was provided by the expansion of oil consumption. The continent as a whole by 1975 was dependent on oil for some 60 per cent of the total energy required (and on natural gas for another 15 per cent). In Africa coal provided about 75 per cent of commercial energy use in 1955; it was down to less than 60 per cent by 1972 and, thereafter, continued to fall steadily. Meanwhile, oil consumption more than tripled between 1956 and 1972 and, moreover, it continued to rise quite rapidly thereafter through the rest of the 1970s in spite of post-1973 economic problems arising in part, out of the much higher price of oil. Out of more than 130 countries and territories in the developing world only five – Taiwan, India, South Korea, Zambia and Zimbabwe – consume more coal than oil.

As has already been hinted, there are many political and economic problems, often difficult to disentangle, connected with the developing world's need for oil. Politically, there is the conflict between the intense nationalism of the developing nations and the foreign control exercised over the oil sector of their economies. In the main, the oil companies have been responsible for the establishment of the complex international infrastructure that gets oil from the points where it is produced, through the refining processes, and then, in the right quantities – generally small – and appropriate qualities to the markets in the developing countries. There is no doubt that they operate efficiently and reliably, and in many developing countries the local affiliates of the major international oil companies present an enclave of efficiency in a generally less than efficient public and private administration. Their employees are usually smartly turned out; their service stations never seem to run out of the required product; their transport is clean and well maintained; they often produce the only readily available maps of the country; and they sometimes finance or sponsor facilities ranging from technical training centres to young people's soccer and baseball competitions. This essentially western image which they present is, paradoxically, sometimes a contributory cause of their difficulties, for to be so wealthy and to be able to 'afford' to be so efficient and such good citizens must mean, according to some of the more vocal of many local populations, that they are 'exploiting' the local economy. The poverty of the many, it is argued, is in part a function of the wealth of the oil companies.

Oil companies operating in such circumstances have often become the 'whipping boys' for the 'colonial' powers which, though no longer having direct political control, certainly exercise significant influence through their control over the purse strings of many developing countries. These charges cannot be completely disproved, though they are probably increasingly exaggerated as time goes by. But certainly in the past to a large degree, and more recently to a smaller degree, the oil companies have had something of a stranglehold over many of the countries in which they have operated. The companies' pricing policies have often been indefensible – though one should note that other nations have also suffered at the hands of the companies' 'posted price and assessed freight rate' pricing system. We saw the impact of this in Western Europe in the immediate post-war period and noted how its disadvantageous effects were overcome, not just by the increasing political strength of European governments as they recovered from the war, but also by the intervention of the U.S. government, which found itself paying too high prices for oil to be delivered to Western Europe by the international oil companies under the terms of the Marshall Plan.

The pricing system, however, lasted longer and worked even more effectively in the interests of the companies in the case of the developing nations. For example, until the middle 1960s any competition for oil products in the markets of West Africa was eliminated in a joint scheme devised by the major companies for supplying the small countries of the region. In this scheme each of the companies involved took its turn at providing a cargo of oil products from one of its refineries elsewhere in the world (e.g. the Caribbean or Western Europe), for delivery to the West African countries. The ship involved moved from port to port along the West African coast, discharging the oil products required into the tanks of all the companies involved. Each company charged – and in turn was charged by others – for the oil it supplied, on the basis of the posted price of the products involved at the Caribbean port of loading, plus the ocean freight rate published for the voyage undertaken. This was done irrespective of what the costs of oil purchase or transportation had been, and irrespective of the price at which the oil could have been obtained from and delivered by a third party. This West African Supply Agreement persisted until some of the West African countries started to refine their own oil. Until then it constituted a small but very profitable business for the companies involved and led to high prices for oil products in the countries concerned.

For India similar penalties had to be paid by the nation and its consumers as a consequence of the country importing both crude oil and products through the international companies. Until the Second World War Indian oil imports were priced as though the oil had been brought out of the United States and shipped from the Gulf of Mexico. In fact, the oil mainly came from the much lower-cost fields of the Persian Gulf, only one-tenth the distance from India, or from the Dutch East Indies, which were less than one-fifth as far from India as the Gulf of Mexico. During the Second World War freight charges were adjusted as a result of pressure on the oil companies, from the British and U.S. military authorities in the Indian subcontinent, to take into account the Persian Gulf origin of the oil. After the war the free-on-board prices of India's oil imports were based on the new Middle East posted prices which, as we have already seen, soon diverged from those of the Gulf of Mexico, where oil production costs were much higher. Thereafter, however, no further adjustments were made to India's oil import prices to reflect the steadily falling costs of shipping oil, as tankers increased in size and scale economies were achieved. Neither was adjustment made by the international companies serving the Indian market, as Middle Eastern oil began to be traded internationally at prices well below those officially posted in the Gulf.

The Indian government was increasingly unhappy about the position during the late 1950s, but it was powerless to take any effective action – other than by persuasion – until it had an alternative source of its necessary oil imports available. Persuasion by the government produced no results, but, then, in 1960 the Indian government was offered oil from the Soviet Union at prices which were over 20 per cent less than the delivered price of similar Middle Eastern oils. Armed with this offer, and having formed a national oil company able to handle imports independently of the international oil companies if necessary, the government told the oil companies which were selling oil in India to reduce their prices. After arguments, threats, and counter-threats the companies eventually did so, though not by the full 22 per cent demanded. Thereafter, on several occasions up to 1971, they had to adjust their prices still further downwards to allow for changing conditions, and particularly to take note of the prices at which the Soviet Union was prepared to offer suitable crude oil and oil products to India.

It is thus only quite recently, and only in the case of important and relatively powerful developing nations like India, that the nations themselves have had any effective voice in the pricing of oil, a commodity which enters into all sectors of their economic life. The developing nations' lack of participation in making decisions which affect their own welfare, and the reservation, instead, of such decisions for outside parties, justified the description of these countries' relationships with the international oil companies as a neo-colonial one. Such relationships have, moreover, increasingly produced similar reactions elsewhere to the one just described in the case of India.

The developing nations' main economic problem over oil has arisen from the impact that rapidly increasing imports of the commodity have had on their balance of payments positions. Most of these countries almost by definition depend upon the exports of one or two primary commodities, and their export earnings have either stagnated or, at best, fluctuated to give good and bad years, as a result of the rapidly changing supply/demand positions for the products concerned. Thus, most of them have only a limited and often a variable amount of foreign exchange available with which to make their energy purchases overseas. But, as we have already seen, these countries' demands for energy have not stagnated or fluctuated in the same way. They have tended instead to slope steeply upwards. Therefore, in almost all developing countries the foreign exchange costs of oil have risen more or less continuously. In some cases, for example Brazil and Argentina, oil import costs came to account for some 25 per cent of the total availability of their total export earnings, even in the period of low international oil prices in the 1960s. Oil needs thus became an important factor making for rigidity in the import structure and, through this, created tendencies towards chronic and permanent balance of payments problems and, often, associated inflationary tendencies in the economies concerned. In the light of this, most of the developing countries had to take steps to try to minimize the cost of oil imports.

One way to reduce the scale of the problem was to initiate a change similar to that which we examined for immediate post-war Europe: that is, the substitution of imported oil products by imported crude oil supplies, with the refining being carried out as a domestic industry. This had the effect of reducing the foreign exchange component in the importation of any given quantity of oil, as crude oil could be bought abroad more cheaply than oil products. The similarities with the European situation should not be taken too far, however, for the developing nations' situation as relatively small oil-using countries did not, in general, create the same enthusiasm amongst the international companies for the local construction of refining capacity. In the case of European countries national wishes in respect of refinery developments usually coincided with the best commercial interests of the companies concerned, because the latter could build economic refineries given the size of the local markets for oil products. Refining, however, had a minimum scale-of-operations requirement of the order of 2 million tons per annum and few individual companies had markets of this size in any of the developing countries in the mid-1950s, when the pressure for such developments really began to mount. There were, of course, exceptions to this as a few of the largest countries, Brazil and India for example, had large enough markets to justify this size of refinery. Even in these areas, however, the markets for products were spread over vast areas. In India, for example, the markets were divided between the main coastal conurbations of Bombay, Madras and Calcutta and several remote inland centres of industry and/or population. This meant that a refinery in one location could not serve all the demand areas at an overall cost which was lower than the alternative of importing products from large refineries in oil-exporting countries, through a number of Indian ports close to the centres of oil demand.

This problem could sometimes be solved by all the companies concerned agreeing to share refining facilities in different localities. They could then arrange exchange deals with each other, so that refineries of the minimum necessary size could then be constructed. In this way refineries in some developing countries were financed by the international companies involved in the markets. In other cases, however, alternative finance for refinery developments had to be sought, either because the companies would not do the job or because the nation concerned did not want them to. The state in these cases sometimes persuaded a third party to come in, or agreed to it coming in. Such a party was often E.N.I., the Italian state oil concern, or an oil organization from one of the communist countries with refinery expertise (the U.S.S.R., Czechoslovakia or Romania). Alternatively, the government of the country could attempt to borrow the money itself in order to construct its own refining facilities.

It might be thought that international borrowing for this purpose would have been enthusiastically supported by the important international lending agencies established in the post-war period to help the developing countries with their capital problems during the process of industrialization. Such agencies have, after all, made a particular point of arranging loans for the construction of hydro-electric plants, roads and other energy and transport facilities which go to make up the economic infrastructure essential for the successful implementation of development plans. Since oil became such a dominant source of energy for most developing countries, it would appear to have been almost self-evident that the financing of refineries would be viewed equally favourably by the international agencies. On the contrary, however, until the late 1960s – that is, until after fifteen years of aid and development programmes – no capital for refinery construction was made available for such purposes, either from the international organizations, such as the World Bank, or within the framework of bilateral aid from western nations, such as the credit extended through the U.S. Aid programme. The reasoning behind this failure was that as private funds were available for building oil refineries, then there was no problem of a capital bottle-neck in this sector. It was far better, so the argument ran, to make the limited amounts of public funds available for projects that could not otherwise be financed at all. Such an argument is valid, of course, pro-viding one accepts that the countries concerned were in a position to negotiate satisfactory financing arrangements with the oil companies – and that the countries accept, unequivocally, the philosophy of the pri-vate financing and ownership of an essential part of their economic in-frastructure. Indeed, oil refineries can be said to represent a 'commanding height' of many of the developing countries' economies. The govern-ments of many such countries did not, however, accept such a philosophy and, in line with an important branch of political thought in Western Europe, where many of the leaders of the developing nations had had their education and where they had also been politically orientated, they considered that oil refining ought to be directly under their own control. Thus the decisions of the World Bank and other institutions not to give loans for developments of this sort were suspected to be the result of pressure from the international oil companies which were anxious to maintain their positions in developing countries. The developing coun-tries saw the denial to them of official loan-funds for refinery con-struction as one of the strategies employed by the oil companies in their role as neo-colonialists.

For countries as large as India, Brazil, Egypt, Argentina and Pakistan the absence of international loans only slightly delayed the build-up of their refining industries, as they employed various devices for securing the capital and expertise required. Sometimes they played off one inter-national company against another, or offered marketing advantages to the company that agreed to build a refinery. Or they invited refinery bids from other parties. Or they built up the oil-refining industry in the public sector of their economy and secured loans to cover the foreign cost component from whatever sources were available. One by one they all managed to become more or less self-sufficient in refined products, except for difficulties arising out of their particular patterns of product demand. This sometimes necessitated additional imports of certain products and sometimes it also meant the international disposal of temporarily surplus products, because it was not economical to have their refineries adjust to such short-term needs. For example, India has usually had to import extra kerosene, which is used extensively for cooking, and to export gasoline, which emerges as a surplus product from its refineries, owing to the relatively small number of cars in the country. Incidentally, the existence of such refinery shortages and surpluses within the framework of a single national market is a factor which strengthens the hand of the international oil companies' proposals when refinery developments in a developing country are under consideration. Such companies can – and do – offer, as part of their proposals, to cope with 'shortages and surpluses' within the framework of their international operations. They thus eliminate the practical problems and high costs which can be involved in an international search for limited quantities of a particular product and/or in trying to sell a product of unknown quality on a world market which may already be overloaded with it, but which product, if it remains unsold, would not only make the refinery unprofitable, but also cause it to work at less than full capacity, so as to avoid the over-production of the unwanted product.

Although, by one means or another, the large developing countries eventually achieved their objective of building up a domestic refining industry sufficient to meet national needs, the same cannot be said of the much larger number of smaller developing countries anxious to achieve the same objective, both for balance of payments and industrialization motives. Their difficulties have been both political and economic. Politically, they had been less well able to build up a state-owned oil industry which could develop refining capabilities. Even when they have been willing and able to do this, or to take other state action to achieve the same objective, they have, nevertheless, been in a less strong position to do so than their larger neighbours. Such smaller countries would not be so likely to receive refinery offers from the communist countries, and even if they did get them and then take action on them they could find themselves at the receiving end of a counteraction by the U.S.A. This happened, for example, in 1964–5 in the case of Ceylon (now Sri Lanka). Ceylon created a state oil-importing and distribution agency, accepted Soviet help and oil supplies to get this working, and was then cut off from all U.S. assistance, as the U.S.A. intervened to retaliate in favour of its oil companies, which considered themselves unjustly treated by the Ceylon government.

From the economic side, most of these smaller developing countries had a total annual demand for oil products of less than 2 million tons, which throughput was, until recently, considered the smallest demand which would justify the construction of a refinery. In such circumstances it remained more profitable for the companies to continue supplying such markets out of their large 'export' refineries in the Caribbean, Western Europe and the Middle East. These had been built with such markets in view, and the companies preferred to use them, rather than to invest capital and managerial expertise in the construction and operation of a series of small refining units. In the case of one Central American country, for example, it was estimated in the early 1960s that the net extra cost involved in manufacturing the country's oil products needs locally in a proposed refinery (which was intended to produce only half a million tons a year) amounted to 60 cents per barrel more – an increase in costs of approximately 25 per cent – compared with the costs involved in supplying the market from Shell's large export refinery, processing 15 million tons a year of Venezuelan crude oil, on the nearby Dutch island of Curaçao.

Knowing that they would have difficulties in ever persuading a government to permit oil product prices to rise by this amount, so as to cover the additional costs involved, and thus appreciating that such local refining would seriously squeeze their profits, the international companies showed little enthusiasm for such developments and used every possible device to stop such developments from going ahead – but all to little avail. As a result of pressure from the governments concerned and, much more important, pressure from companies prepared to make such refinery deals as a means of getting access to markets in which they hitherto had no interests, refineries have been built in such countries. By the beginning of the 1970s few such countries remained without a refinery.

In Central America, for example, the first refinery was opened in 1963 on the Caribbean coast of Guatemala. Since then one refinery has been built in each of the other Central American countries and Guatemala even secured a second one on its Pacific coast. The total oil product demand in the five countries was, and still is, under 6 million tons per annum. This is little more than the throughput of what was hitherto considered to be a reasonably sized economic refinery, yet the seven Central American refineries, with an average capacity of only just over a million tons per year, offer a total capacity of over 8 million tons per year. Thus, not only are they high-cost plants by virtue of their small size, but their unit costs of production also suffer from their partially unused facilities. It has been calculated that there is a 1 per cent rise in the unit costs of refining for every 1 per cent of under-used capacity, so the economic penalty of the existence of surplus facilities should not be underestimated. Thus, for the companies this new pattern of providing Central America with its oil requirements appeared to be anything but a lowest-cost method. For the Central American countries, while the refineries led to some foreign exchange savings and created a relatively small number of new jobs in the industry, they also put upward pressure on oil prices. There was, without doubt, some fat to be skimmed from the profits that the companies were previously making in selling oil products to their countries. Much of this disappeared with the development of the refineries and the consequent upward movement of prices was, of course, to the ultimate disadvantage of local consumers. This was particularly dangerous for the largest oil consumers in the countries concerned (companies, that is, that were generally concerned with the basic export sectors such as banana production). They had previously been able to buy their oil requirements cheaply in the open international market (by tender), but they now became obliged, at government insistence, to buy from the local refinery, at prices which were necessarily higher than those to which such enterprises were accustomed. The continuation of this kind of requirement sometimes helped to make the exporters' operations uncompetitive on the world markets, and thus led to a decline in their activities, to the obvious detriment of the countries' economies.

In such cases it is impossible to argue that the establishment of refineries was justified. The countries concerned could have saved just as much foreign exchange, and would have enabled prices to oil consumers to be reduced as well, by insisting that the companies brought down the import prices of their products from the levels of 'posted prices' to the price levels of an open market. The example of Western Europe and of the largest developing countries in encouraging domestic refineries was not necessarily one which ought to have been followed by the smaller developing countries, such as those in Central America. Refinery development would better have waited until the time when their demands for petroleum products justified the construction of a large enough refinery. The disadvantages of development prior to this may well have outweighed the advantages.

For many developing nations, however, not even the political and economic benefits flowing (or considered to flow) from the establishment of a domestic refining industry were sufficient to meet their aspirations. The development of indigenous crude oil resources, whereby oil imports can be substituted entirely, seemed to offer even better prospects for additional foreign exchange savings and other economic benefits, plus more effective political independence.

The less risky, but often also a less politically satisfying, way of achieving oil production has been to persuade one or more of the international or other foreign companies to conduct a geological and geophysical survey which, if favourable, could then lead to their undertaking exploration and development work. Evidence for the companies' recognition of the powerful potential of such nationalistic feeling can be seen in their willingness to do this on so many occasions in the 1960s; particularly when one remembers that this coincided with a period in which most companies have had control of, or access to, more than enough crude oil from the main oil-producing areas of the world to make further exploitation not only unnecessary, but also an apparent misuse of their capital resources. One explanation for this apparently irrational behaviour could be their altruism, but this is unlikely, given the increasingly hard battle that the companies had to maintain their overall levels of profit at an acceptable height in the 1960s.

There are two other possible explanations. One is that individual companies agreed to do such work in order to keep other companies and institutions out of a particular country with oil potential, thus minimizing the danger that oil would be discovered by other entities that might consequently set themselves up as rivals in the local oil market. Until quite recently in most countries the right to prospect and explore for oil could be bought very cheaply, and some companies have from time to time undoubtedly secured acreage simply to eliminate the possibility of its exploration by other potentially interested parties. For example, some of the international companies which had oil concession rights in India and Pakistan were alleged to have undertaken their searches and developments with less than sufficient enthusiasm, diligence and urgency, because they preferred to continue to supply these countries with oil from their vast, low-cost reserves in the Middle East. Such charges are difficult to prove, but from the companies' point of view what they were alleged to have done made commercial sense, even though they denied the allegations. In any case, however, more and more countries came to demand guarantees of minimum exploration and development programmes from any concessions that were awarded, and so the opportunities for 'sitting' on possible sources of oil merely to keep out competing companies, or to ensure the continuation of a market for an existing investment in oil-producing capacity in another country, diminished quickly.

The second explanation for the willingness of companies to undertake searches in new areas arises from the clear indication given to them by the governments concerned that any oil discovered at home would be given absolute priority over imported oil. In these circumstances, foreign companies with investments in marketing and perhaps refining facilities in such a country have had a very positive incentive to find domestic oil – before anyone else did so. Without its own local supplies a company

179

would, in such circumstances, be obliged to refine and market someone else's domestically produced crude oil. To make the bad situation even worse, such obligations would probably be within the framework of a system of price controls which all but eliminated the profits from local refining and marketing operations, but which still left the possibility of making good profits out of the domestic oil production.

An excellent example of this occurred in the late 1950s in Argentina, when foreign companies were invited to explore for and/or develop oil resources. The invitation was accompanied by firm promises that any oil found would be guaranteed a market in Argentina, that the companies concerned would be paid a price for their oil which was at least as high as the delivered price of crude oil from overseas, and that such domestic oil supplies would immediately replace imported supplies as far as possible. In this situation Shell and Esso, the two international oil companies which already had local Argentinian interests, could not afford not to participate in the search for indigenous oil. They had built up important interests in marketing and refining oil in Argentina, as a result of which their parent international companies were making considerable profits out of selling them crude oil from Venezuela and the Middle East at delivered prices equivalent to posted prices in the Caribbean or the Persian Gulf, plus the full freight rates allowed by the quoted assessments (which were generally much above actual costs). If they found oil as a result of their searches in Argentina then, given the Argentine government's willingness to pay import parity prices for the oil, there seemed every possibility that they would make just as good, or even better, profits out of this arrangement than from the pre-existing one. If, on the other hand, they chose not to search for local oil, but to argue their preference for continuing to bring in supplies from overseas (an argument unlikely to be acceptable to the Argentine government), and another company found oil, then they would be obliged to take it in their refineries, instead of being allowed to continue to import it from their overseas affiliates. Thus, all their profits on crude oil supplies would be lost. Such policies of economic nationalism in many of the world's developing countries in the 1950s and the 1960s played the most important role in persuading the international oil companies to extend and diversify their areas of exploration. There were thus relatively few countries of the world (outside the communist countries) which did not at some time in that period manage to get some kind of an oil survey and development programme under way, even where the geological and other physical prospects may not have been too good. Unhappily, in the circumstances of readily available low-cost international oil, such activities seldom led to the development of production.

Not all developing countries have been prepared to accept the economic and/or political implications of having foreign companies come and look for oil. Such countries chose instead to establish a national oil company with exploration and production responsibilities. In many ways this represented an extension of the 'commanding height' philosophy as far as oil was concerned, in recognition of the need for the national ownership of industries which are of fundamental importance to all or most sectors of the economy. In Latin America, most of which has been politically independent for more than a century, such views have been commonplace for many years, and state participation in Argentina, Chile and Bolivia dates as far back as the 1920s. In 1938 Mexico's reaction against the oil companies went much further when it expropriated all the assets of the petroleum companies operating there and established PEMEX – Petróleos Mexicanos – with a complete monopoly over all oil industry activities in the country, from exploration to marketing. More recent examples of this attitude are seen in Peru's nationalization late in 1968 of the producing and other operations of the International Petroleum Company – a subsidiary of Esso – and Bolivia's expropriation of Gulf Oil's producing and export operations in 1969.

These well-formulated attitudes relate to the fact that the companies involved have been concerned with Latin American oil from the earliest days of the industry – and over this long period they achieved a widespread reputation locally of being 'agents of economic imperialism'. This reputation made the companies susceptible to attack from a wide spectrum of political groups within the Latin American countries, for they thus became involved in the major political issues of nationalism and its associated anti-Americanism of more recent years. In such a situation one can readily understand the reason for the great propensity on the part of Latin American nations to keep the development of their oil resources in their own hands.

In Chile the state monopoly over oil exploration and production, first established in 1927, was maintained until the late 1970s during which period E.N.A.P., the state company, gradually expanded its producing operations in the extreme south of the country (in Tierra del Fuego) with the aim of making Chile self-sufficient in oil. The best it ever achieved was the production of a little less than two-thirds of the country's total requirement, and it was even having difficulty in maintaining this share in view of a steadily expanding demand for oil by the industrializing economy, and the apparent relative poverty of Chile's oil resources. Exploration in the northern parts of the country was unsuccessful and the southern fields which had been found seemed, until 1977, to be capable of only modest expansion. Since then interesting new discoveries have opened up the possibilities of significant increases in the levels of production, though the capital needs and other problems associated with such a development in a very remote area remain to be solved. These problems were, indeed, the reason why the international oil companies were eventually invited back to Chile to help with the exploitation process.

In Brazil, too, oil exploration and production was a state monopoly from the formation of Petróbras in 1950 until 1980. As a result of extensive and expensive searches Petróbras discovered oilfields in Bahia, which have been producing oil in gradually increasing quantity since 1954, and there are now developments under way to secure additional output from parts of the Amazon basin and from newly discovered offshore fields. But Brazil's oil development to date, like Chile's, has remained limited relative to growing national oil use and it is only in recent years, with the development of the country's newly discovered offshore resources, that Petróbras' level of production exceeded one half of the country's requirement.

E.N.A.P. and Petróbras argue, however, that without their efforts their countries would not have been producing any oil at all, because the international companies in the international oil market conditions of the 1950s, the 1960s and the early 1970s would not have found local efforts to be worthwhile, and would have opted instead to supply all of Chile's and Brazil's needs from their major supply sources in Venezuela and/or the Middle East. The companies argue a different case, saying that had they been given opportunities to search for oil they would certainly have done so. They argue, moreover, that with their greater expertise and resources they would most likely have done better than the state entities, whose relatively small quantities of oil production from a necessarily limited exploration effort have been achieved only at the cost of capital investment which could have found more worthwhile and less risky outlets in other sectors of the capital-short economies of the countries concerned.

To back up their arguments the companies cite the case of nearby Argentina, where the state oil company's monopoly over production was ended in 1958. This was because of the country's growing burden of oil import costs, the state company's long-term inability to raise oil production to more than a third of the country's total needs, and the unlikelihood that it would ever be able to do much better, in spite of the known existence of potentially highly attractive oil areas. This was basically due to the inability of the Argentine exchequer to provide the state company with the capital it required to do the job. Foreign companies were thus invited to participate in the search for oil. With their ready availability of capital and their managerial and technical expertise they were able, within three years of accepting the invitation, to increase production threefold and thus make Argentina self-sufficient in oil. This technical success appears to have been bought, however, at the cost of a considerable loss of foreign exchange to Argentina, as the successful companies concerned with the oil exploration and production programme repatriated their profits out of the country. This repatriation, moreover, was at such a high level, a later government claimed, that it more than offset the foreign exchange savings which were achieved by the replacement of foreign crude by domestic supplies. These views were challenged, but the validity of the claims and counter-claims on the economic front proved difficult to substantiate.

What is beyond doubt, however, is the fact that President Frondizi's decision to relinquish the state oil monopoly in 1958 was the major cause of his overthrow a couple of years later. The government which replaced him soon reintroduced the state monopoly over oil, through Y.P.F. (which incidentally soon allowed production once again to fall below consumption). Even more recently the situation was again reversed, as a military government in the country decided that the foreign companies were necessary to produce the oil the country required for its development. The formidable increases in the price of oil between 1973 and 1982 underlined the economic importance of exploiting indigenous oil resources, even if this necessarily involved arrangements with the international oil companies in which, indeed, most of the world's expertise is still to be found. In recognition of this fact of life even Chile and Brazil, as indicated above, joined Argentina in making such arrangements in the early 1980s as a means of minimizing the foreign exchange costs of imported oil which was then available only at prices of up to $40 a barrel. We shall return to the question of the contemporary position of the developing countries in the world of oil power later in the chapter.

Looking at Chile, Brazil and Argentina it seems that the argument in Latin America between those favouring state development on the one hand and those favouring foreign oil company development on the other has been 'won on points' by the latter. But this conclusion can only be reached if one ignores the success of Mexico in producing a state oil entity – PEMEX – which achieved the basic objective of providing the country with the oil it needed in the process of development through industrialization. Though one may argue over the wisdom of certain aspects of PEMEX's activities, and though one recognizes that it had a flying start, as a result of the availability of oilfields as 'going concerns' at the time of the expropriation, it is still true that PEMEX demonstrates that foreign companies are not essential to ensure that domestically produced oil becomes available to a developing nation anxious to eliminate an import bill for foreign oil of growing dimensions. Indeed, PEMEX has recently taken its contribution well beyond this with its discovery of very large new oil reserves in southern Mexico. These have enabled Mexico to become an oil exporter of increasing significance since 1980 (see above, Chapter 3, for a discussion of Mexico's oil potential) and with prospects for continuing increases in export levels for the rest of the decade.

Thus, in spite of the warnings (from the U.S.A. and international lending agencies such as the World Bank, as well as by the international oil companies) against spending scarce domestic capital on oil exploration and development, more and more developing countries have set up national entities run by the state to pursue the search for oil within their national territories. To date, however, except in the Latin American countries mentioned above, such entities have made very little contribution to their nations' oil supplies.

The reasons for this are not difficult to find. Oil is an elusive commodity and the major requirement for finding it is capital. It is, moreover, high-risk capital as it involves investment in an enterprise which may not produce any returns owing to the absence of economically exploitable resources. Very few of the developing countries have sufficient capital to spare for such risk ventures on a large enough scale. Thus, an alternative strategy has been evolved by a few developing countries in order to overcome not only the foreign exchange problems arising from their need to continue to import oil, but also their political dislike for their oil suppliers. They argue that the international companies really act as little more than middlemen in getting oil from producing to consuming areas and, in doing so, 'exploit' both producers and consumers. Why then should producers and consumers not get together in joint ventures to their mutual advantage? This argument led India, for example, to seek to obtain supplies directly from producing nations in the Middle East and in 1967 the first agreement along these lines was signed with Iran. One field was brought into production, but in spite of the eagerness of India to secure these supplies, which were negotiated to be made available at a lower foreign exchange cost than that involved under the pre-existing arrangements for importing oil, little of the oil involved moved to India, as it proved to be unsuitable for India's existing refineries. On the other side of the world, in Latin America, attention has been given over a number of years to the integration of Venezuela's ability to produce crude oil and the rest of South America's need for it, but so far little effective action has materialized. Brazil's state oil company – Petróbras – has also created joint venture operations with the state entities in Iraq and Algeria. In the case of Iraq, Brazilian enterprise successfully found a major oilfield and its exploitation initially gave Brazil a guaranteed supply of oil. Unfortunately, Iraq felt that the nationalization of this field was also a necessary part of its post-1974 oil policy. More recently, Petróbras's joint ventures with the Iranian national oil corporation have similarly run into difficulties as a result of changes in Iranian oil policy.

Government-to-government discussions between countries of the Third World have been based on the idea that oil should flow from one developing country to another without the intervention of the international oil companies. Such possibilities became much more interesting after 1973 with the rising importance of O.P.E.C. and the nationalization of discovered oil reserves by many of the oil-exporting countries. Thus, new patterns of relationships between O.P.E.C. countries and oil-importing developing countries began to be investigated. The emergence of political awareness and realism in the countries of the developing world, and the creation of instruments of economic policy whereby they can help each other, could, in the longer term, help to eliminate the role of the international oil companies in serving them. This will, however, still take time and, meanwhile, the oil-consuming developing world will continue to have to rely on American and European companies to a high degree, with the inevitable accompaniment of suspicion and hard feelings between them as a fact of international life. This has particularly been the case over the last decade, when the companies have had to pass on to the poor importing countries the higher oil prices which followed the revised agreements between the oil-producing countries and the oil companies. The fact that the companies initially made these agreements collectively, and then seemed to raise prices by more than the additional tax burden justified, was seen as evidence that their exploitation of the poor oil-consuming countries was increasing. As a result Brazil, for example, immediately announced its intention to make direct government-to-government arrangements for its oil supplies, as well as its intention to redouble its efforts to produce more oil at home.

This sort of reaction seemed likely to become very common in the 1970s amongst the countries in the developing world which depended upon imported oil for their economic advance, but their faith in the wisdom of such government-to-government deals was soon rudely shattered by the traumatic events in the oil world which followed the establishment of oil-producing countries' control over the supply and price of oil. The poor oil-importing countries quickly found that the 'lucky few' major oil-producing countries in the Third World appeared to have even less concern for their economies and their prospects for economic development than had been the case with the oil companies and their owning nations (the United States, Britain and the Netherlands). The latter may have supported the oil companies' efforts to refine, distribute and market oil profitably in the developing oil-importing nations, but the profits sought and earned were modest (of the order of $1 to $2 per barrel at most), compared with the very much higher returns which the oil-exporting countries started to earn out of sales to the developing world in the aftermath of the quadrupled price of their oil in 1974. Revenues for the oil-producing countries of $10 per barrel or more (on a commodity costing 10–30 cents to produce) became the expected norm, in spite of the fact that the landed price for each barrel of oil multiplied by the number of barrels needed by the developing countries gave a total foreign exchange cost for the commodity which, as shown in Table 2 (below), many of them at that time could not afford to pay, either from their annual export earnings, or from their foreign exchange reserves.

Table 2. Impact of increased crude oil prices in 1974 (in millions of dollars) on selected oil-importing countries in the Third World

Country	Foreign trade 1973		Balance of trade	Additional cost of oil in 1974*	Foreign exchange reserves (end 1973)
	Exports	Imports			
Ethiopia	285	180	+ 105	45	114
India	2,934	2,771	+ 163	1,050	629
Pakistan	983	928	+ 55	220	254
Philippines	1,494	1,243	+ 251	550	606
Sierra Leone	145	125	+ 20	22	36
Sudan	401	376	+ 25	105	28
Tanzania	399	363	+ 36	45	54
Thailand	1,382	1,662	− 280	525	1,107

* On the basis of prices for crude oil in 1974 and with no relief from these prices for the developing countries. Most developing countries' balance of trade positions deteriorated from their generally favourable position in 1973 as the changes in the international oil system caused the demand for other primary products to fall away as a result of recession in industrialized economies.

This already adverse situation for the oil-importing developing countries was then severely exacerbated by the second oil-price shock of 1979–81. Thus, in order to buy the oil they needed to enable their development even to have the possibility of continuing to go ahead, they have had to go into debt. Collectively, their debt by 1982 grew to over $300 milliard ($10^9$), such that many of them reached the virtual limit of their borrowing abilities. Since then their indebtedness has further increased to an estimated level in mid-1985 of over $1,000 milliard. On average the cost of servicing the debts involved now accounts for well over half of the total foreign exchange earnings of the oil-importing developing countries.

Pleas by the poor oil-importing countries for help from the oil-exporting countries were initially ignored or rejected, except in special cases (such as Iranian help for Pakistan). Later, some oil exporters offered low-cost loans to importing countries to enable the latter to buy the oil they needed. Such offers, which have since been much extended, help in the short-term, but they do nothing to assist the longer-term problem of the poor oil-importing countries, and it seemed essential that some adjustment to this gross imbalance would need to be sought. This could, perhaps, have been through the creation of a differentiated pricing system, whereby the oil producers charged higher prices to the industrialized world and lower ones to the poor countries. This sort of action would certainly enhance support for the O.P.E.C. countries amongst other Third World countries. That, or an alternative sort of action with similar effects, such as the joint help extended by Venezuela and Mexico through the San José agreement to the small oil-importing countries of the Caribbean and Central America, would be a politically astute move for the oil producers to make as a means of enhancing their position in the world of oil power. O.P.E.C. initiated consideration of these issues as long ago as 1978, following the presentation of a proposal for such action by the President of Venezuela at the O.P.E.C. meeting held in Caracas in December 1977. Ways of making some significant and systematic help available to the world's poor countries, in compensation for the high cost of the oil they need to import to continue the development process, has to date, however, remained a matter for serious discussion by O.P.E.C., rather than a matter on which collective action has been agreed and taken. Action has been left principally to individual members of the organization, as shown above in the case of Venezuela.

Even the poorest oil-importing countries are unlikely ever to get their oil imports at anything like the $3 per barrel price they did in the 1960s. They may therefore have to accept that, in respect of oil at least, they have something in common with the rich importing countries. And, as in the cases of United States, Western Europe and Japan, the oil-importing countries in the developing world will have to give much more attention and devote many more resources over the next decade to the development of whatever indigenous sources of energy they happen to have – such as coal in India, oil shale in Thailand or hydro-electricity in Brazil. Even more urgent and more important, because it offers the chance of producing short-term positive results, is the intensification of the search for conventional oil at home. Recent studies by international organizations such as the World Bank and United Nations specialized agencies, as well as by the geological services of both the United States and the U.S.S.R., have all shown that there is a potential for oil and gas production on a scale large enough to satisfy national needs in a large number of developing countries which currently have to import most of the energy they use. There are financial and institutional problems which currently inhibit the efforts needed in this respect and further attention will be given to these in Chapter 9, as their solution seems likely to constitute a main component in changing the world of oil power in the last fifteen years of the century. From the point of view of the poor oil-importing countries concerned the discovery and exploitation of their reserves of conventional oil and gas is now of greater importance than ever before. The burden of oil imports has become so great, and the uncertainty created by dependence on imported oil so high, that, without a successful effort to substitute oil imports by indigenous supplies (of oil and gas) in the short term, the economic development prospects for many countries of the Third World seem unlikely to be good enough to satisfy the needs of their growing populations over the coming years. This would, of course, not only be disastrous for the countries and peoples concerned, but would also add a further element of danger to the already much more dangerous world of the post-1973 revolution in the world of oil power.

8. Oil in International Relations and World Economic Development

The preceding chapters of this book have attempted to describe and evaluate the role and significance of oil in different nations and groups of nations. The contrasting, and in some ways conflicting, interests of the U.S.A., the main oil-producing nations, the principal consuming countries of Western Europe and Japan, the developing nations with their rapidly increasing demand for oil, and the U.S.S.R. and its allies have been stressed. In addition, it was necessary to take into account the growing complexity of the global oil industry's organization, for the companies and other institutions such as O.P.E.C. concerned with oil have legal, economic and political relationships with the world's nations and inevitably form part of the international oil system.

Little more than a decade ago there were fewer than a dozen of these entities which really mattered at an international level, and to a large degree they did what seemed best from their own point of view. They took note mainly of each other and they worked together where possible (as in the joint producing companies in many countries of the Middle East). Where this was impossible, usually because of U.S. anti-trust legislation, they rarely moved in such a way as to upset the others and thus displayed an 'understanding' which, to the casual observer, perhaps appeared to indicate the continued existence of the cartel arrangement of the 1930s between the seven international oil companies. Today, quite apart from national, and in some cases, regional governments, there are at least 150 entities which form part of the oil system. Such large numbers inhibit formal agreements and render complete understanding between them well-nigh impossible, even when use is made of mathematical and sociological models and techniques to enable 'predictions' to be made of others' responses to a particular course of action by one of the entities or groups of entities involved.

But the world oil industry is more than the interrelationship of supply sources and markets, of companies with governments, and of companies with companies. The system works within the framework of the larger and even more complex system of the world's political and economic relationships, and this book would be sadly incomplete if there were no attempt to outline the main points at which the 'oil system' impinges on, affects and is affected by the larger 'international system'. This chapter cannot be exhaustive in its treatment of these issues. Instead, we shall choose a few cases with which to illustrate some of the general points involved.

At a purely political level the interrelationships of oil and the Middle East cauldron of instability and change have made the headlines on more than a few occasions since the Second World War. The political stresses and tensions in the area have certainly had an impact on the oil industry in the countries of the Middle East (as shown in Chapter 4), but in return the development of the latter has certainly produced a different Middle East from that which would otherwise have emerged.

The tensions in the Middle East arising from the Arab–Israeli conflict have, time and time again, produced their effects on the oil industry. The initial formation of the state of Israel was immediately followed by the cutting of the pipelines which had previously taken oil from Iraq to Haifa for refining or onward transhipment to Western Europe (see Map 4, pp. 90–91). To this day these pipelines remain broken and unused and the Iraq Petroleum Company was obliged to replace the sections in Israel and Jordan with lines through Syria and the Lebanon, where new refining and export terminals were built at Tripoli.

More recently these replacement pipelines have also been closed from time to time, both by the political problems in the Lebanon and by Syrian reactions to Iraqi policies. Iraq, indeed, now sees its most secure crude oil outlet to the Mediterranean as the pipeline to Turkey. The capacity of this line, originally built in the mid-1970s, has recently been increased so that Iraq could become less dependent on the state of its relationships with Syria for the continuity of its oil exports. Israeli territory also had to be avoided when Aramco, a few years later, decided to build the Trans-Arabian Pipe (Tapline) from its fields on the Gulf coast of Saudi Arabia to the Mediterranean coast. Supply to a terminal in Israel would have been the cheapest way of providing this transportation, but political conditions made this impossible and so an extra thirty miles or so of line had to be laid around Israel's northern boundary. Since the Arab–Israeli war of June 1967 even a few miles of this relocated line have been in Israeli-occupied parts of Syria, but so far Israel has not interfered with its operations though, inexplicably, Arab commandos have done so. Their action temporarily closed the line at one stage and also interrupted the transit revenues earned by Lebanon, Syria and Jordan. Partly as a reaction to the insecurity of its oil export system arising from the repeated conflicts in the Levant, Saudi Arabia has built a new crude oil export line across its own territory to a new port on its Red Sea coast. The shipment of oil from there to the markets of Europe is subject only to the policies of Egypt with its control over the Suez Canal. Egypt's peace treaty with Israel perhaps persuaded Saudi Arabia that this route for its oil exports was likely to be a more secure one. Its attractiveness is emphasized by the fact that other Gulf oil producers are now linking their oil fields into this line in order to avoid tanker transport problems arising from Iran's threat to close the Straits of Hormuz (see Map 4, pp. 90–91). Indeed, Iraq's link is already operational since it had to be completed as quickly as possible once Iran's military successes along its border with Iraq closed off the latter's access to the Gulf. More recently, however, Iraq's air attacks have severely curtailed Iran's use of its main crude oil export terminal on Kharg Island so that Iran too has decided to construct an alternative pipeline system to take its oil overland to a port in southern Iran beyond the reach of Iraqi planes and outside the disputed waters of the Gulf.

From the very beginning of its conflict with the Arab world in 1947 Israel has been denied Arab oil and for many years the Haifa refinery, designed originally to run on Iraqi crude, stood idle. Israel's oil needs, first as oil products, and then as crude oil once the Haifa refinery had been reconstructed to take oil of other specifications, have had to be imported from other, more distant parts of the world, such as Venezuela and more recently Mexico. Even then it could be received only in tankers which avoided the use of the Suez Canal, which until its re-opening in 1975 (after its long period of closure since 1967) was forbidden by Egypt to ships carrying goods to and from Israel. The international companies which marketed oil in Israel were also forced by the Arabs to choose between continuing to do such business in Israel and being allowed to pursue their interests in Arab countries. As the importance of the latter invariably exceeded the importance of the former to the international oil companies, Israel had to organize its own refining, distribution and marketing facilities. It thus achieved the virtually unique position in the non-communist world of being unable to 'Go Well with Shell', or to share the advantages of having 'Tigers in the Tanks' of its motoring public!

But Israel's oil problems arising from the continued state of war with the Arab countries have at least been predictable and thus amenable to solution, albeit at a cost of higher transport and other charges. Less predictable, however, have been the indirect effects on the oil industry of Israel–Arab tensions since 1947. In times of the absence of active and large-scale hostilities the oil business could continue working, provided the Arabs' ground rules forbidding contact with Israel were observed; but when the continuous tension deteriorated from time to time into actual warfare, the significance of the Arab–Israeli dispute for the oil industry became virtually universal. In both the Suez War of 1956 and the Six Day War of 1967 not only was the Suez Canal closed, with a consequent traumatic effect on the international movement of oil, but the Arab states also tried to use their control over a large part of the world's oil production as a weapon to assist them in their struggle. In 1956–7, the situation arising from the closure of the Suez Canal, the sabotage of some of the pipelines from producing countries to the Mediterranean, and the ban on shipping Arab oil to Britain, France and certain other countries produced several serious consequences. First, it necessitated the short-term rationing of oil products in Western Europe. Second, it immediately led to increased oil prices as the companies sought to cover the increased costs arising from their need to buy more oil from more expensive areas (notably the U.S.A.), and from the longer tanker journeys involved in shipping Middle Eastern oil round the Cape of Good Hope. And third, it produced intense diplomatic pressure on the part of the U.S.A. and other western nations to get Egypt and other Arab countries to desist in their use of oil as a political weapon. This just succeeded and the Western European economies escaped major close-downs by a hair's breadth.

Ten years later, in 1967, Middle Eastern oil was even more important to most of the world's industrial economies. As explained earlier low oil prices had encouraged its replacement of coal to such an extent that oil had become the dominant industrial fuel in most countries (see Chapters 5, 6 and 7). Moreover, motor transport had grown rapidly in importance in these countries since the mid-1950s and the transport sector of the economies concerned had become largely dependent on oil products. On this occasion, therefore, it seemed that a repeat of the Arab nations' 1956 actions in using oil as a weapon in their fight against Israel must cause a major breakdown in the western world's economic system. At first, the pattern of 1956–7 in this respect appeared to be repeating itself. The Suez Canal was once again closed and reports of bans on oil movements to certain countries were announced by governments and oil workers of producing nations. This time, however, there was no sabotage of the Mediterranean pipelines and the embargoes on shipping oil to particular destinations were, in general, lifted after a few days. The world's unused tanker capacity (which existed partly because of the over-ordering of tankers in the aftermath of the Suez crisis of 1956–7) was quickly chartered to move the oil round the Cape of Good Hope.

This route was, in fact, already being taken by the largest oil tankers which had been brought into service in the 1960s in the search for economies of scale in oil transportation as they were unable to proceed fully laden through the Canal. Moreover, in the decade since the previous crisis new oil-producing capacity had been developed in parts of the world from which the movement of oil to Western Europe and the United States was not dependent upon the Suez Canal – notably Libya, Nigeria and Algeria. Additionally, other areas of relatively low-cost production – for example, Venezuela – were working at less than their full capacity. In 1967, therefore, Western Europe was not denied the much increased volume of oil it then required and the only penalty it had to pay for the renewed outbreak of hostilities in the Middle East was temporary higher prices. These were needed in part to cover the additional transport charges but, in greater part, they reflected the reaction in the oil market to the short-term imbalance between supply and demand.

The continued closure of the Suez Canal between 1967 and 1975 certainly altered the relative abilities of different countries' oil (and of different companies for that matter) to 'compete' in the world markets. In particular it worked to the advantage of Venezuela, Libya and Nigeria, whose oil does not have to go through the Canal to reach Europe. Nevertheless, the Middle East's oil production continued to grow and it is somewhat ironic that the single most important interruption to the established patterns of international oil supply arising from the Canal's closure was the U.S.S.R.'s inability to send oil from the Black Sea to Japan. Fortunately for the U.S.S.R., whose oil trade with Japan constituted an important source of foreign exchange earning, it was possible to arrange a 'swop deal' with B.P. This willingness on the part of B.P. to come to an arrangement with the Soviet Union reflected the fact that B.P. as a company was particularly badly affected by the Canal's closure, because of the dominance of Middle Eastern supplies in its total oil availability. Under the terms of this agreement oil was taken out of Iraq by B.P. for shipment to Japan and was 'exchanged' on a ton-for-ton basis with Soviet oil delivered from Black Sea and Baltic Sea export terminals to B.P. outlets in Western Europe. This arrangement continued even after the re-opening of the Canal, for all the parties concerned benefited considerably from the overall reduction in transport costs incurred.

Whether or not the Suez Canal was ever again opened for oil tankers after about 1970 became almost a matter of indifference to many of the oil companies concerned. Minor adjustments to their international production patterns, coupled with their decisions to achieve, as quickly as possible, the ability to move oil in tankers whose size and speed made it less costly for them to transport crude oil to Western Europe or the U.S.A. via the Cape of Good Hope rather than through the Canal, marked an important turning point in the long history of the Canal as a primary route for the international movement of oil. And if this development in tanker technology was not sufficient to seal the fate of the Suez Canal as the main link between the world's most important oil-exporting region in the Gulf and the world's main markets for imported oil in Western Europe and North America, then the new multi-million-ton-per-year pipelines designed to carry oil from the Red Sea to the Mediterranean certainly seemed likely to do so (see Map 4, pp. 90–91). One of these is the Israeli pipeline which was built from Eilat, on the country's narrow coastline on the Red Sea, to Tel Aviv on the Mediterranean coast. This development became possible when Israel secured control of the Gulf of Aqaba following its occupation of the Sinai peninsula in 1973. Given an Arab world which accepted Israel and Israel's right to take advantage of its location to provide transhipment facilities for Middle Eastern oil en route to Europe, then this pipeline would have made very good sense indeed. But since the Arab world did not, and in general still does not, accept these propositions, then the economic viability of the pipeline has always been in doubt. None of the Arab oil-producing countries ever agreed to their oil being transported through the line. The only country that made use of it for moving its oil exports in large quantities was Iran under the Shah. Even then the throughput of Iranian oil was limited as most oil from Iran continued to be produced and shipped until 1978 by companies that also had large interests in the Arab states. Thus, they could not risk getting overtly involved in the Israeli project because of the threat this would constitute to their interests in Arab countries. Since 1978, moreover, the revolutionary government in Iran has changed the Shah's policy of cooperation with Israel. Indeed, Iran has become even more vociferous and active in its opposition to Israel than most Arab states and there is thus now no question of the Israeli pipeline being used (other than surreptitiously) for moving oil from Iran.

The only remaining large-scale possibilities for its use, other than for moving oil to Israel itself, have always been and remain quite remote. One possibility was the north–south movement of Soviet oil exports bound for India, Japan and other Far Eastern markets, but this would have meant such a fundamental switch in Soviet political strategy in the Middle East, away from its support for the Arab states, that it was a highly unlikely development indeed. A second possibility could have been the south–north movement of Australian and other Far Eastern oil exports to Western Europe by companies new to the international oil business, and thus without interests in the Arab countries to consider. This development depended on a large-scale build-up in supplies from such new oil-exporting areas and on securing markets for these supplies in European countries, where the large international oil companies were well established. Neither of these was a likely occurrence so that such new international oil business could not be relied upon to provide the traffic to justify a large, and hence low-cost, trans-Israel pipeline. Therefore, the likelihood of a major expansion in its capacity, or even the full use of its existing capacity, remains small, unless and until there is a comprehensive peace treaty between Israel and the Arab nations.

The only exception to this conclusion would have been if Iran, on taking over complete control of its oil industry and thus retaining the ownership of much more oil in transit to centres of consumption than had hitherto been the case, had wanted to ignore entirely Arab feelings over the question and so take full advantage of possible economies to be gained in moving its oil via this pipeline to Western and Eastern Europe. Under the new regime in Iran this, as indicated above, has turned out not to be the case. In any event, Iran does have a potential transport alternative to this politically somewhat dangerous use of Israel's line should there be any need, because of deteriorating political relationships with the Arab states of the Gulf in general and, in particular, with Saudi Arabia, with which it shares control over the Straits of Hormuz, to supplement its use of low-cost tanker transportation of its crude oil exports from its Gulf terminals to Western and Southern Europe. This alternative is the proposal for a very large diameter crude oil pipeline to take Iranian oil production directly from the oilfields in the south-eastern part of the country to the Mediterranean coast of Turkey. As can be seen from Map 4 (pp. 90–91) this would not only have the effect of reducing by about 75 per cent the distance over which the oil has to travel to reach the Mediterranean (compared with the use of the Israeli line), but it also avoids crossing any country other than Turkey. This potential development would eleminate the only opportunity which remains for Israel to become an important crude oil transit country. Now, of course, this not only depends on Iran's post-revolutionary government changing its policy towards Israel, but also on Iran managing to find enough markets for its oil in Western Europe to justify the large investment required. The prospects for either condition being satisfied are now relatively remote.

The second Middle East pipeline which has been built is largely, though not entirely, free from the political problems that have affected the use of the Israeli pipeline. This is the Egyptian development of twin 100 cm diameter lines from Suez to Alexandria capable of transporting 80 million tons of oil a year (and which could be increased to 125 million with the addition of more pumping stations). It has, moreover, been built with facilities for loading and unloading the largest tankers in use at both the Red Sea and the Mediterranean ends of the line. Its construction can be interpreted in economic terms as Egypt's response to the loss of business by the Suez Canal when it became cheaper for large oil tankers to use the Cape route. It might, however, have been built even without this development in world shipping, as an alternative to widening and deepening the Canal itself in order to provide extra transit capacity for additional oil traffic. The Suez pipeline is, of course, subject to the same dangers as the Suez Canal in the event of hostilities involving Egypt. Nevertheless, the fact that the line is served, at both ends, by large tankers which could in the event of such trouble also use the alternative Cape route, led the oil companies to decide to use its facilities. It gives an opportunity for cost-savings which are great enough to offset the relatively limited amount of capital they needed to invest in terminal and storage facilities. However, given the continuing tension between Israel and the Arab world, and in spite of the peace treaty and the establishment of special relations between Egypt and Israel, no major oil company has been prepared to put all its eggs into this particular basket. The Suez pipeline has thus, to date, been able to do little more for Egypt than to enable it to retain some part of the oil transit traffic from which, in the absence of super-tankers and political hostilities in the Middle East, it could have expected to secure increasing revenues by providing a route for moving oil from the Middle East to Western Europe and North America. Since 1984, however, the economic viability of the Suez pipeline has been significantly enhanced by Saudi Arabia's construction of a crude oil pipeline to its Red Sea coast at Yanbu (see Map 4, pp. 90–91). The decision of Saudi Arabia to export some of its oil to Western Europe via Yanbu only made economic sense because of the existence of the Suez pipeline. The subsequent development of hostilities in the Gulf – as part of the Iran/Iraq War – has enhanced the importance of the new Saudi pipeline to Yanbu and its capacity is being increased. Moreover, as indicated above, other Gulf states and Iraq itself are also using – or planning to use – the same route for exports of their oil so that Egypt's Suez pipelines will benefit to an even greater extent – though given the decline in demand for Middle East oil in Western Europe (see Chapter 5) by less than would otherwise have been the case.

Egypt's keenness on pressing ahead with this pipeline project, which was completed in 1976, provided a further piece of evidence to support the view that the Arab states after 1967 began seriously to re-evaluate the use of their control over the industrialized world's oil requirements as a weapon in their geopolitical struggle against Israel and its western allies. In 1956–7 there were unilateral decisions on the part of Arab governments, and/or by particular elements in their countries, to close down oil production and oil transport facilities in order to bring pressure to bear on the western oil-importing countries, both directly, and also through the effects of such decisions on the commercial interests of the international oil companies. From the 1960s onwards, however, the Arab oil-exporting countries achieved better returns from the oil companies' operations so that their own benefits from oil increased very rapidly. Their immediate and longer-term economic and political viability thus came to depend increasingly on keeping their oil flowing.

On the other hand, the threat to use oil as a political weapon had persuaded some consumers of Middle East oil to try to decrease their dependence on it, as they were not content merely to await the next unilateral decision by the Arab states to turn off their essential energy requirements. As previously shown, the U.S. decision in 1959 to introduce mandatory quotas on crude oil imports was taken partly because of concern for the security of supplies from the Middle East. Japan, with a greater dependence on Middle Eastern oil than any other major industrial nation, has, as shown in Chapter 6, encouraged national companies to develop oil resources elsewhere, and it persuaded other companies supplying oil to the Japanese market to diversify their sources of supply. The countries of Western Europe, which collectively provide the largest market for Middle East oil, and which at the time of the Suez War depended on it for 75 per cent of their total oil supplies, also reacted to some degree to the dangers and risks of too rapid a change to imported oil as the basis of their energy economies. Without the fear of interruptions to Middle Eastern supplies, most Western European countries would have had even fewer inhibitions about pursuing cheap-energy policies based on imported oil than was the case between 1958 and 1970. Instead, indigenous coal was given some degree of protection, while alternative supplies of imported energy, including coal from the U.S.A. and oil from the U.S.S.R., were accepted as a means of reducing the risks. Still more recently, the risks attendant upon a continuing high degree of dependence on Middle Eastern oil have also helped to enhance

Western Europe's interest in its own natural gas and oil potential. In fact, the potential of North Sea and Dutch gas, plus good prospects for additional gas supplies from the Adriatic Sea and other offshore areas, plus the development of the newly discovered indigenous oil resources, already as early as 1970 seemed likely to be able to eliminate by the end of the decade much of the additional annual need for imported oil in Western Europe. In the light of the agreements in 1971–2 between the oil-producing countries and the international oil companies over the supply and price of oil up to 1981, it was generally felt in Western Europe prior to 1973 that there was little real danger of a reduction in Middle Eastern supplies of oil in the 1970s but, thereafter, the dangers would increase so that serious attention should be given to alternative supplies.

These attitudes and developments in the oil-consuming world did not go unnoticed in the Middle East, even though in the media they received little publicity, compared with references to the industrialized world's dependence on Middle East oil. This realism helped to account for the relatively limited support, in terms of oil sanctions and other interruptions to oil supplies, given to the Arab cause in the 1967 war. It was not simply coincidence that only two new producers – Libya and Algeria, which were less sophisticated in the oil business than the longer-term producers of the Middle East proper – persisted in boycotting certain markets for more than a few days. In Saudi Arabia, Kuwait and Iraq, as well as in the smaller and newer producers of the sheikdoms of the lower Persian Gulf, no action was permitted which would have meant a long-term interruption of supplies. Embargoes were restricted to certain destinations and certain ships (depending on nationality) and no really serious steps appear to have been taken to ensure that the companies did not get away merely with observing the letter, but not the spirit, of the boycott decisions. The companies reorganized their supply arrangements so that all bans on particular oils to particular destinations by particular ships could be observed, but they switched tankers and cargoes so that no nation went short of oil (and, in doing so, incidentally, helped to demonstrate the validity of their claim that they can act as neutral intermediaries in moving oil between hostile governments). In other words, the Middle East oil-producing nations appeared by 1967 to have made up their minds that, in both the short and the longer term, they themselves were likely to suffer the most from simply 'holding the west to ransom' over oil. The closure of the Suez Canal – and Egypt's reluctance to allow it to be cleared and re-opened for years after the cessation of hostilities – was beyond the control of the Arab oil-producing countries.

Given the speed and degree to which oil transport was reorganized, so that the closure of the Canal came to matter very little and certainly to have no effects other than very temporary ones on the oil exporters' revenues, the financial support the oil-rich countries gave to Egypt (see below) to compensate it for its loss of oil-transit revenues could not be described as support for the use of the 'oil weapon'.

Thus, after the 1967 war, at a series of Arab conferences held to determine future policy, a new philosophy towards the use of oil as a means of achieving desired ends began to emerge. In brief, the new policy aimed to make the deliberate use of revenues earned from exporting oil the means of providing the financial base for the achievement of political ends. The scope for this new policy, emerging not only from the growth in total oil exports but also from the growth in revenues per barrel (they rose from about 30 cents per barrel in 1946 to about $1 by 1970), is clear. In 1972 the Middle East (including Libya and Algeria, but excluding Iran) exported over 5,000 million barrels of oil and secured some $5,000 million in revenues. The first small indications of the possibilities which existed for using oil revenues for political ends came with the small monthly payments of approximately $5 million which were made by two of the major Arab oil-producing nations (Kuwait and Saudi Arabia) to Egypt while the Suez Canal was closed, and to Jordan, for an indefinite period, to help with the costs of the Palestine refugees and the loss of its West Bank to Israel. These initial steps in the use of oil revenues for supporting political objectives of the Arab oil exporters have been much developed in the period since 1973. The very much higher revenues available to the Arab oil-producing countries have been used in part for supporting a series of causes considered worthwhile. (These are discussed in Chapter 9.)

There thus appeared to be recognition in the Arab world as early as the late 1960s that oil revenues could be used as a main instrument for achieving the geo-political aims of the Arab group of states. Such a development had a twofold significance in international relations terms. First, the flow of oil from the Arab exporting countries would be unaffected by political crises if the producing countries were to attach more importance to ensuring the continuation of revenues from oil than to using oil as a political weapon. This has happened to a large extent in later Middle East disturbances (with the exception of the use of the oil weapon for political purposes by Iran, a non-Arab country, in 1978–9). The development did not, of course, as has been clearly demonstrated since 1973, exclude the use by the exporting countries of oil as an 'economic weapon' against the rest of the world, but this was a general response by the oil-producing countries to their hitherto very limited role in the international system, rather than one related specifically to Arab producers in a Middle East context. It is discussed fully in Chapter 9.

Second, as a result of the economic and political advantages that oil revenues could buy, there might well develop an Arab Middle East with an enhanced internal cohesion and a more significant geopolitical potential among the power blocs of the world. This was achieved in large degree after 1973, when it became somewhat ironic to find the commodity which originally helped to divide the region into the spheres of influence of competing outside powers (and which later emphasized national rivalries within the region by creating 'have' and 'have not' nations in the Arab world), emerging to provide the means whereby greater regional cohesion and strength between the Arab nations became closer to reality, albeit through the 'use' of an unfriendly, but a commercially necessary, outside world. In institutional terms this has been expressed in the Organization of Arab Petroleum Exporting Countries (O.A.P.E.C.). This has proved to be one of the more successful international bodies in the Arab world – not only in terms of the degree of cohesion it has given to oil and oil-related policy-making amongst its members, but also in developing contacts and relations with other parts of the world – most notably in respect of financial and technical aid to developing countries for petroleum and petrochemical projects.

The presentation so far of the relationships between oil and geopolitics in the Middle East has been restricted to a consideration of the impact of the Arab–Israeli conflict. Important though this is, it is not the only geo-political consideration arising from oil questions and problems in an area which has been a centre of world interest and conflict for many centuries. The limited concern we can give here to other considerations relates to an evaluation of the significance of oil in interesting outside powers in the region and in influencing their policies. The earliest outside interest in the region's oil came from Britain and Germany, with the latter at first more active through its influence in the pre-First World War Turkish empire. Its activities were not, however, very successful when compared with the results of exploration by British interests in Persia. Success in those exploration efforts provided more than enough oil for Britain, whose political control over the rest of the Persian Gulf area could then be used to inhibit possible oil developments by others. Germany's interests in the region were, of course, eliminated by its defeat in 1918, when the division of the former Turkish empire into British and French spheres of control led to a general understanding between these two countries to divide the 'spoils' as far as the prospects for oil were concerned.

Thus, the newly formed state of Iraq, where the prospects for oil appeared good, lay on the frontier of British and French influence and this necessitated the participation of both British and French capital in a consortium which was created to initiate oil exploration. West of Iraq, the former Turkish empire lay wholly within the French sphere but the successor states there – Syria and Lebanon – turned out to have little by way of oil production prospects though, at a later period, they offered routes and export points for the pipelines running to the Mediterranean from the producing areas further east. As a result French oil interests in the region turned out to be quite limited and France's disappointment in this respect had an important influence on the country's negative attitude to the international oil industry. Britain's success in Persia on the other hand enabled it gradually to extend its oil interests to other parts of the Persian Gulf, where its earlier informal political authority was formally confirmed in 1919.

In the inter-war period British companies discovered oil resources along the Gulf coast, particularly in Kuwait. The opportunities for marketing Middle Eastern oil at that time were not great enough, however, to give a sufficient incentive to develop these resources. The existing fields in Iran and Iraq were then quite capable of producing whatever oil could be sold out of the Middle East, the oil potential of which was, indeed, effectively inhibited by the world oil supply and pricing system, dominated by the U.S.A., and the existence, unofficially before 1933 and officially after that, of the international oil cartel. (See Chapter 1.) These attributes of the international oil industry in the inter-war period enabled the U.S.A., Mexico and, to a lesser degree, Venezuela and the Dutch East Indies to maintain their dominant role in world oil markets and so restrained the expansion of known Middle East resources. But as the prolific nature of the Middle East oilfields and the low cost of extracting oil from them, compared with production costs in the United States, was gradually appreciated, so the U.S.A. showed increasing interest in the region. Thus, the political efforts of Washington were linked with the commercial and technical efforts of the American oil companies in a joint attempt to gain access to the oil wealth of the region. This involved a diplomatic struggle with Britain and France, as well as rivalry between oil companies of the different nations involved. Eventually, as a result of its intense political and commercial pressures, the U.S.A. secured an entry to Middle East oil-producing areas. The British government managed to keep American interests out of its most important producing and refining facilities in Persia, but it was forced to allow them into Iraq and the Persian Gulf states. In Iraq, the consortium, in which British and French companies retained an interest, came to be dominated by a group of U.S. companies, while in Kuwait the concessionary company became fifty-fifty British and American. Thus, by the outbreak of the Second World War in 1939, the U.S.A. had already won for itself a position of virtual parity with Britain, and a much stronger role than France, in spite of the latter's longer-standing oil interests in the region.

During the early years of the war the British military presence throughout the area was greatly strengthened and was used to secure what were considered to be vital British interests. Once the U.S.A. entered the war, however, its greater reserves of personnel and equipment quickly gave it a military influence of little less significance. Meanwhile, the defeat of France in 1940 more or less eliminated French influence in the Middle East, with Syria and Lebanon, the former French dependencies, quickly achieving independence. By 1945 Britain and the U.S.A. had thus 'tied up' the region in political terms, and so paved the way for their oil companies to move in to take advantage of the concession areas which they had secured within the framework of this colonial situation. Internally, Britain and America now faced the task of keeping the area under their political control. As oil became more important, ran the British and American argument, so more effective control became necessary. Political control could not, however, be maintained after 1950 within the framework of the traditional colonial relationship, and the two outside powers were gradually obliged to acquiesce in the increasingly vocal and active demands for independence. Subsequent demands from the newly independent oil-producing countries also led to the previously described renegotiations of the oil agreements, to give the countries concerned much increased financial returns from, and even some control over, the activities of the oil companies. It was the combination of political and economic moves towards independence for the Middle Eastern countries which gave rise to concern, not only in Britain and the U.S.A. but also in all the other countries which depended on the region's oil, about the security of supplies. There has thus now been a period of over thirty-five years of uncertainty concerning oil supplies from the Middle East, dating from the Iranian nationalization of its oil industry in 1951 and marked from time to time by what usually proved to be temporary and partial disruptions of the oil flow. Similarly, disruptions to supplies have also occurred as a result of the breakdown of negotiations over prices and other issues between the producing countries and the companies holding the concessions. This set of issues eventually culminated in the traumatic events of 1973 and 1974, whereby the relative power of oil companies and oil governments over supplies and prices was fundamentally altered. This development is the subject of the next chapter.

Anglo-American efforts first to secure, and then to maintain, direct or indirect control over Middle East countries were only partly motivated by their desire to ensure the 'rights' of their companies to explore for and develop Middle East oil, and to guarantee its transhipment to world markets. In part, they were dictated by the post-war fear that the Middle East might become increasingly susceptible to political intervention by the U.S.S.R. which, it was suspected, not only had a political interest in the Middle Eastern countries, but also an economic interest in securing control over the region's oil. Some observers noted the rapidly increasing demand for energy within the Soviet Union and assumed that this demand could not be met by production within the country's boundaries. They also recalled the very early post-war efforts by the U.S.S.R. (up to 1948) to secure concessions for oil exploration and development in the northern parts of Iran, which they had militarily controlled, by agreement with Britain, from 1941. It was therefore predicted that the U.S.S.R. would seek access to Middle East oil and would do this not on commercial terms, through purchases from the producing companies (or even countries), but within the framework of a political effort to secure the support of the growing Arab nationalist movement. In the face of this threat, the U.S. and the U.K. governments forgot their earlier differences on the division of the Middle East's oil wealth. Instead, they emphasized the need to present a united front in the region against the common enemy, and attempted to sell to the Middle East countries the idea of the peripheral anti-Soviet alliance, as originally developed in Western Europe through NATO. It was from these efforts that the Central Treaty Organization (CENTO) emerged, in which one of the most important oil-producing countries of the Middle East, Iran, as well as Turkey and Pakistan, decided to participate. Behind the 'screen' thus provided by these three Middle East countries, Anglo-American interests were felt to be more secure, and both diplomats and oil investors breathed a little more easily once the Treaty had been signed in 1955 (see Chapter 2).

With hindsight one can see that this concern about Soviet oil intentions in the Middle East was exaggerated, for the U.S.S.R. proved to have more than enough oil and gas within its own borders to meet its rapidly-growing requirements. In fact, the rapid development of the Soviet oil industry enabled the country not only to change from a coal-based to an oil- and gas-based economy but also, from the late 1950s, to become something of a rival to the Middle East in respect of oil exports (see Chapter 3). Middle Eastern producing countries and companies now faced the challenge of a new competitor and one, moreover, which was prepared to cut the prices of its exports of oil and gas in order to break into the markets controlled by companies with large-scale production in the Middle East. It would, indeed, appear to be the very success of the U.S.S.R. in so developing its international contacts that enabled it to conclude some commercial arrangements with Middle Eastern countries over oil and gas.

It concluded an agreement, for example, to purchase large quantities of Iranian natural gas (via a pipeline link between the two countries). This Iranian gas is nearer to Soviet consuming areas than some of its own unutilized resources of Siberia, and hence available at lower cost. It also enabled the U.S.S.R. to use the Iranian gas to replace both domestic oil and natural gas, so making the latter available for export, and able to earn more foreign exchange than the Iranian gas cost the U.S.S.R. The Soviet Union also agreed to help with marketing the oil produced from those parts of Iraq exploited by the latter's new state oil company, to which financial and technical help has also been extended by the Soviet Union. This offer also provided a new basis for the kind of international exchange arrangements described earlier, whereby the costs of supplying U.S.S.R. export markets in the Far East are reduced by taking the oil requirements from Middle Eastern countries and then using the Soviet supplies themselves in the closer and more accessible European markets. In view of the continued high level of Soviet oil production and, even more significantly, its still rapidly increasing gas production, linked with the discoveries of a succession of very large fields in the vast areas of potentially petroliferous sedimentary basins of the U.S.S.R., there remains little chance that the Soviet Union itself will in the foreseeable future have to rely other than marginally on the Middle East for its essential energy supplies. This does not exclude the certainty that the Soviet Union will continue to attempt to exploit political opportunities in the Middle East for extending its influence there, as in the 1980 takeover in Afghanistan, or in Iran in the aftermath of the revolution there, or in the context of the Iraq–Iran war. The motivation for such action in the short to medium term will not, however, be to secure access to energy supplies which it must have for its economic development, in the same way that Western Europe, Japan and even the United States to some degree came to depend on such supplies from the Middle East by 1973.

The one remaining motive for Soviet intervention in the region lies in a deliberate attempt by the U.S.S.R. to gain control over Middle Eastern oil supplies, in order to deny them to those countries that depend on them. Such a deliberate policy would, however, be regarded as tantamount to an act of war against the west and hence likely to bring the confrontation which neither the U.S.S.R. nor the U.S.A. seek or desire. Such a dangerous policy by the Soviet Union would appear to be out of keeping with the realities of major power relations in the 1980s and in any case it would hardly be worth the risks involved. There is now no guarantee that the denial of one, two or even three Middle Eastern countries' oil would cause much more than a temporary supply problem for the western world, where the long-term outlook for energy still appears to be one of potential surplus rather than of shortage, in spite of appearances to the contrary from time to time in the aftermath of the changes in the world of oil power since 1973. Though a little more time is still needed for the west to switch from what are perceived to be the unreliable supplies of O.P.E.C. oil to other sources of energy supply, this has been happening faster than the Middle East suppliers expected, so that the demand for their oil is much less than they expected. Thus, a policy of Soviet intervention in the Middle East designed to deny oil to western consuming nations would now certainly not be welcome to the Arab nations concerned. They have a continuing need for revenues from oil to be sold to the west in order to provide them with the foundations for their political and economic strength. They would be very unhappy with a Soviet attempt to eliminate their exports as this would destroy the basis of their power.

It is thus in the Middle East that the interrelationship of oil and international politics is at its most significant and its most complex. Elsewhere their interrelationship is usually a bilateral one, most often between the national interests of a particular country and the interests of the U.S.A., as a result of the latter's ownership of the majority of the major oil companies that operate in many parts of the world. These bilateral difficulties arising from the importance of oil and from its organization internationally have, sometimes, tended to escalate to multilateral proportions as, for example, in the case of Latin America, where an almost continent-wide consensus exists over attitudes to oil and the reactions to the largely American companies concerned with its production and distribution.

Individually over the last forty years, or even longer in some cases, the countries of Latin America have viewed the international oil industry as a particularly effective manifestation of U.S. imperialism. Over this time, as we showed in Chapter 7, one Latin American country after another fought its own national battle with the international oil companies. These battles have usually ended either in complete expropriation and an entirely nationally owned industry, as in Mexico in 1938, Cuba in 1961 and Venezuela in 1976, or in the establishment of powerful national companies such as E.N.A.P. in Chile or Petróbras in Brazil. These entities were then given a dominant role in the national oil scene, while the international companies were relegated to providing supplies of crude oil, services and know-how, but without any real control over the way in which the industry is run. So strong have national feelings been over this issue that governments have sometimes been brought down because they have made or have seemed likely to make too great concessions to the international companies. Perón was overthrown in Argentina in 1955, after ten years of popularity in office, when a deal with an American oil company appeared likely to thwart popular feeling on the subject. Frondizi, another President who followed shortly after Perón and who when elected to office pledged to maintain the *status quo* over oil, fell even more quickly after he implemented steps to bring in foreign companies to help Argentina to find oil. More recently, President Belaúnde Terry of Peru was overthrown because of the agreement he made with Esso over the working of the latter's long-discovered reserves in the country (see also Chapter 7).

But now such bilateral conflicts are being 'multilateralized' through the cooperation of the state oil companies within Latin America, and the attempts to evolve a common petroleum policy within the framework of free-trade areas and/or common markets within the continent. To date, the state entities and the civil servants involved have done little more than talk and draw up statements of intent, or sign proposals for cooperation which could not possibly be implemented. But sooner or later, and more especially since 1973, given the much enhanced motivation to take rather more effective action about oil as a result of the massive price rises for the commodity, the Latin American organization that has been formed with headquarters in Peru, Asistencia Recíproca Petrolera Estatal Latinoamericana (ARPEL), could eventually take positive steps towards assuming control and direction over intra-Latin American trade in crude oil and petroleum products, and over the determination of patterns of production and refining in the region, with all that such action implies in respect of intervention in investment decisions, which the international companies have up to now so jealously guarded.

Already, some advance in this direction has been seen in the case of the European Common Market. This is slowly moving towards a multinational agreement on energy policy that could involve a common policy towards the international oil companies. The latter will be constrained to work within the new regional framework laid down by the political authority and will have to justify their attitudes and actions not only to the governments of the individual countries which make up the Community, but also to the Common Market authorities as well. Again, moves in this direction were greatly stimulated by the new difficulties and dangers of the international oil scene and were described in Chapter 5.

In both Latin America and Europe there are also examples of the way in which the potential for oil discovery and development necessitates agreements or causes disagreements, over the division of the continental shelves, the exploration of which has become a major new frontier in the oil industry's expansion. The best example of a political agreement in this respect was the division of the North Sea between Britain, Norway, Denmark, the Netherlands and West Germany (see Map 7, pp. 137–39) so that each country was then able to enact national legislation to regulate the exploration and exploitation of its part of the North Sea, an essential prerequisite for concession systems to be determined for the oil companies to start work. Similarly, such median-line approaches to, or other methods of, arranging the division of the continental shelf to the west of the British Isles between the United Kingdom and Ireland, and between the U.K. and France, in respect of the English Channel and the Western Approaches, have been agreed though some details remain to be finalized. By way of contrast, at the other end of Europe a potentially dangerous dispute has flared up between Greece and Turkey, because the two countries have not been able to agree on a division of the Aegean Sea between them for purposes of offshore exploration in that highly potential area for oil occurrence.

Meanwhile, in Latin America an agreement on areas of offshore oil interest has been made between Venezuela and the Netherlands Antilles, lying some thirty to fifty miles off the north coast of Venezuela. On the other hand, Venezuela and Colombia have still to reach a final agreement as to how the Gulf of Venezuela shall be divided for oil exploration purposes. However, the most important disputes in Latin America over offshore oil potential exist in the extreme south of the continent. One which persisted for many years, until it was finally solved in 1983 following arbitration by the Vatican, was between Chile and Argentina over the division of the islands and waters to the south of Tierra del Fuego. A second dispute which still persists is in respect of the very extensive continental shelf to the east of Patagonia. This is highly 'prospective' from the oil and gas standpoint and it is claimed in its entirety by

Argentina. However the Falkland Islands, a British colony over which Argentina also claims sovereignty, lies on the shelf so that the possibility of oil wealth under this part of the South Atlantic complicates the long and recently hard-fought dispute between Britain and Argentina. The invasion of the Falkland Islands by Argentina and their recapture by Britain in the early months of 1982 have done nothing to solve the issue: on the contrary, relationships have been exacerbated and the prospect for a solution, so that oil exploration of the continental shelf could start, is now further away than before.

It is thus in respect of oil questions that issues relating to the Law of the Seas have recently been most effectively tested. Interestingly, questions of territorial rights and jurisdiction over coastal/continental shelf waters have also arisen in respect of some Federal states. Louisiana and Texas challenged U.S. Federal rights in the Gulf of Mexico. In neighbouring Canada, Newfoundland, a province with adjacent offshore waters which include areas of proven oil potential, claims that these are part of the province, rather than part of the Federal offshore regions. Given the rapid growth that is now taking place in continental-shelf and continental-slope activities, disputes over – and, sometimes, agreements on – their political ownership are becoming more important. Perhaps the most testing case in this respect will be related to the question of the East Asian continental shelf, over which China seems likely to lay full claim in respect of mineral exploration rights, to the consternation of many neighbouring territories and western Pacific islands, including even the Philippines, the nearest island of which is over 600 km from the coast of China. Even further in the future, but potentially too important to be ignored, is the question of the ownership and rights over the marine regions beyond the continental slopes. Some of these are not without their oil and gas potential, and disputes caused by the wealth-generating capacity of the oil industry may well extend sometime later in the century to the deep oceans of the world. Oil questions are thus included amongst the most important issues which have necessitated the recent intensive and long-lasting international discussions on the Law of the Seas.

Finally, one finds that oil has been involved even in the world's growing concern with racial problems – at least in so far as these have been reflected in the rest of the world's relationships with South Africa and, until recently, Zimbabwe (formerly Rhodesia). The Union of South Africa is still deliberately pursuing policies of racial segregation and of built-in inequalities in its multi-racial society. Whether or not such policies are a domestic matter of national interest only is not our concern here. We are concerned with the fact that the international community has chosen to intervene over racial policies in particular countries, and that oil has been chosen as an instrument of that intervention.

In the case of the dispute between Britain and Rhodesia, over the illegal declaration of independence by the latter from the former in 1965, the first economic sanction applied by Britain was to prohibit the supply of oil to the rebel regime. Support for oil sanctions was sought specifically amongst the world's oil-producing and transporting nations (as well as generally through the United Nations), and British naval units were stationed off the Portuguese East Africa port of Beira (through which oil for Rhodesia was received for onward transportation by pipeline to the refinery at Umtali) to prevent tankers from completing their journey. Had all Rhodesia's neighbouring states and all the companies involved in supplying oil to the country been committed to the policy of denying oil to Rhodesia, Britain's action would also have made life unpleasant – if not impossible – for Rhodesia's white inhabitants, and it would soon have been successful in crippling the country's economy. This could have been achieved in spite of the fact that Rhodesia was less dependent on oil than almost any other country in the world (except for those with still largely subsistence economies) as a result of the local availability of high-quality, low-cost coal which still provides the basic fuel for industry and the railways and, together with hydro-electricity from the massive new station at Kariba on the Zambezi, for the production of electricity. Rhodesia, however, had neighbouring states which refused to support oil sanctions. Thus, both South Africa and Portugal allowed the British, American and French oil companies which marketed oil in Rhodesia, and which also declined to observe the sanctions requirements of the British, American and French governments, to make oil products available to Rhodesia across their national frontiers. The countries and the companies between them thus limited the impact of oil sanctions to little more than nuisance value. The nuisance was nothing more than the somewhat higher prices which had to be paid by Rhodesian consumers, in order to meet the higher transport costs of getting the oil by road or rail from the neighbouring countries, and the temporary rationing of petrol for use in private cars. The latter, however, proved to be necessary only in the very early days of sanctions, when alternative supply arrangements were still being organized by the companies in cooperation with the Rhodesian authorities, and after January 1974, when Rhodesian supplies through South Africa were affected by the oil boycott imposed on the latter country by the most important Arab oil producers in the aftermath of the October war with Israel.

The effect of this boycott was, as it turned out, relatively short-lived and Rhodesia was able to return to the *status quo ante* the boycott, with the oil supplies it needed coming in via its neighbouring countries. In 1976, however, the situation again deteriorated with the demise of Portuguese rule in Mozambique and the succession to power there of a radical government which basically cut off all relations with Rhodesia. As a result Rhodesia became dependent entirely on the flow of oil products through South Africa which, though still prepared to help, nevertheless now insisted, in the aftermath of the Arab countries' embargo on supplies of oil to it and of the increasingly strong U.N. resolutions on the situation in Rhodesia, on even more convoluted ways for disguising the fact that oil products were being supplied from the South African refineries of the international companies to its northern neighbour. South Africa was by this time, as we shall see below, becoming increasingly concerned about its own oil supplies and thus did not wish to court trouble by openly enabling Rhodesia to avoid the embargo. As a consequence supplies of oil to Rhodesia became tighter and this contributed to the weakening of the Rhodesian economy. Controls on the use of oil had to be re-introduced in respect of civilian use, so that the Rhodesian armed forces, which were involved in a struggle against African guerrillas based in neighbouring countries, could be ensured the oil they required.

The overall effectiveness of oil sanctions against Rhodesia was, however, never as great as had been expected, partly because of the ability and willingness of the oil companies concerned to continue to make their international facilities for acquiring and transporting the oil available to serve Rhodesia, and partly because the sanctions did not extend to South Africa. The possibility of such an extension has never been far from the South African government's fears over the last decade, and the prospect has long been given serious consideration in the country's economic and strategic planning – not unwisely in the light of repeated calls for such action against it from many parts of the world. South Africa, however, as with Rhodesia, also has cheap coal in abundance (cheap, incidentally, partly because of the availability of low-cost African labour). The coal industry was, moreover, even in the days of cheap oil, given additional protection against possible competition by its location on the inland plateau where most of the country's energy consumption occurs. Coal thus has to stand only limited transport costs in being moved to consumers – in marked contrast with the high transport charges which have to be met to get oil from the ports of import to the main demand centres. Moreover, these costs have been kept higher than would otherwise have been the case by the government's insistence, until the early 1970s, that this oil traffic be moved by the state-owned railways.

After much pressure from the oil companies a pipeline from Durban to Johannesburg was authorized, but only when the companies agreed to accept the participation of the railways in the project – a device by the government which not only aimed at maintaining railway revenues, but also one which would have the effect of increasing the charges above the minimum possible, and so maintain a greater degree of protection for South African coal production.

In spite of the availability of low-cost coal, South African oil consumption has, however, increased rapidly, particularly in those uses where oil products have no rivals. Thus, the vulnerability of the economy to international sanctions over oil has also increased steadily. Government recognition of this has been accompanied by action designed to minimize and contain its effects. Of most significance in this respect is the government's long and effective sponsorship of the oil-from-coal plants at Sasolburg. Nowhere else in the non-communist world has the large-scale production of oil from coal yet been economically possible in peacetime conditions. Even in South Africa, with its low-cost coal and the relatively high delivered prices of oil products, the development of such facilities has only been made possible because of the government's willingness to provide the subsidy and the political support required, in return for the guarantee that the oil-from-coal plant provides a minimum availability of oil products for military and other strategically necessary uses. The increasing seriousness with which the government evaluated the risks to the country's oil supply was demonstrated by its decision in 1968 that the oil-from-coal plants should be more than doubled in capacity, with a view to the eventual production of about 30 per cent of South Africa's petrol and diesel oil needs. These expanded facilities are now on stream and, meanwhile, the government has decided to initiate an even larger-scale expansion at a cost which was initially reported to be more than $1,000 million. This new plant is now nearing completion, at a cost which has greatly exceeded the initial estimates, but which brings to almost 50 per cent South Africa's own production of its petroleum products needs. The South African government has also required the massive build-up of stocks of crude oil and oil products in an effort to improve its short-term bargaining power in any crisis. It has in part been prepared to finance these itself (particularly stocks of products required for possible military use), but it also required an undertaking from the oil companies concerned with marketing oil in South Africa that they would hold large stocks at their own expense. This requirement was, indeed, made one condition of the franchise given to oil companies to build or expand refineries. The size of South African oil stocks is a state secret – as is all information concerning oil movements to and from and within the country – but they are thought to provide at least two years' supplies of essential oil needs in the event of a cessation of oil supplies from overseas.

For the longer term the South African government has followed two lines of action to meet the threat. Firstly, it has encouraged and sponsored petroleum exploration in the far from favourable geological conditions of the Union, most of which is composed of geological provinces in which oil is not likely to be found. In the limited areas of oil potential however – such as the Karoos immediately behind Cape Town and in South West Africa – exploration has been pushed ahead, though still with very little success. Encouragement has also been given by the government to persuade the companies to look at the country's rather difficult and narrow continental shelf. As a result, exploration in several offshore areas got under way, with results which were more promising. Drilling in 1982 indicated a gasfield which has been declared commercial (though small) and an oil discovery. A continued search for oil and gas is thus still considered to be well worth while.

Oil production on any scale in South Africa is thus still a matter for conjecture – and, on the part of the South Africans, for hope. South Africa did, at one stage in the 1960s, take encouragement from the recent expansion of oil-production prospects in Angola, where significant offshore oilfields have been found. This then seemed certain to lead to the production of relatively large amounts of oil surplus to the requirements of Portugal and Portuguese African territories and this, it was thought, could find its way by coastal or even overland connections to South African refineries, without fear of intervention by the outside, anti-South African world. These discoveries in Portuguese territories, coupled with the oil sanctions against Rhodesia and the intense South African interest in the problem of its oil needs, constituted the principal arguments in favour of the formation of a southern African bloc in which mutual assistance over oil would be a main area for agreement. This would have helped to minimize the future degree of economic control that the outside world could exercise over these countries and territories by using oil as a weapon. The possibilities in this direction were, however, soon undermined by the independence secured first by Angola in 1974, then by Mozambique in 1976, and finally, by Zimbabwe (Rhodesia) in 1980. This has left South Africa effectively isolated in the southern part of the continent and with a need, therefore, to look to other parts of the world for countries which will be prepared to help it over its oil-supply problems.

In this respect South Africa still does not see the world overseas as unanimous in its attitude towards it over oil and this led it actively to pursue appropriate external policies. Thus, there has been a diplomatic 'offensive' designed to secure the willingness of other nations to continue to supply it with oil in the face of 'world opinion'. Much of this activity has, of necessity, been conducted in secret, but what first emerged to public view as a result of these efforts was an agreement with Iran in the late 1960s. Under this agreement Iran agreed to supply the crude which was to go to a new South African refinery to be partly Iranian owned. This meant, of course, that Iran had a significant national interest in continuing to allow oil to be loaded for South African destinations and, until the overthrow of the Shah in 1979, it resisted all international efforts to get it to impose sanctions against South Africa which thus came to depend on Iran for most of its oil. Since then the new régime in Iran has become militantly anti-South African and no Iranian oil is now officially allowed to go to South Africa. Instead, South Africa has to 'play the international market' to get its supplies and has had to resort to subterfuge to ensure that oil is shipped to its oil import ports. In spite of the difficulties and the publicity generated by revelations on which companies' and countries' ships are trading with South Africa, there is little evidence that the country has not normally been able to get all the oil it wants, albeit at a premium price: including significant amounts from Arab states, in spite of the boycott declared by the Organization of Arab Petroleum Exporting Countries (O.A.P.E.C.) and from Iran.

The South African government also tried to secure guarantees on the continuity of supplies from the foreign oil companies with refining and marketing interests in South Africa. Indeed, it seems highly likely that all of them have given an assurance that they will do what they can should the occasion arise. French companies, in particular, appear to have satisfied the South African government in this respect to a greater degree than have the American or British ones, and they were 'rewarded' with an interest in new refinery capacity which was built in South Africa.

Finally, South Africa has recognized that the minimization of the oil risk also provides an economic case for pursuing 'southern hemisphere' solidarity. Since the late 1960s South African diplomats, particularly those concerned with oil, have been active in South America, and there are signs that they believe the contacts they have made could lead to an oil flow, should it turn out to be required. Indications are that Argentina was considered to be the possible source of supply as it had a government until 1982 which was more favourably inclined towards South Africa than most others in the world. This pointed in the direction of a possible future involvement of South African interests in the very considerable Argentine oil potential and, in this respect, South Africa must have had a keener interest than most other countries in seeing that the continental shelf off Patagonia was worked by Argentina. Looking in the other direction around the southern hemisphere, South Africa has also been very encouraged by the significant finds of oil and gas in Australia and New Zealand in recent years, for South African markets could be among the most attractive of any in the world for any production which Australasia has surplus to its own requirements, though such exports are unlikely to be permitted by the present governments of Australia and New Zealand. In the light of these significant preparations which South Africa is making against all eventualities, and in the light of the continued willingness of a number of the major international oil companies (British, French and American) to treat South Africa as an integral part of their international logistical systems, it seems unlikely that the use of oil sanctions, as a means of action against it in respect of its racial policies, will be any more successful than were the wartime efforts of the allied nations between 1941 and 1945 against Germany, when even that country's military machine managed to secure the fuel that it required.

Alternative stratagems are always available to a nation to secure essential oil. There are, for example, usually high profits to be made out of sanction- or blockade-breaking by individuals or individual companies prepared to take the risks involved. And, in any case, it is very unlikely that unanimity of action on such an issue by all interested parties can ever be achieved. Thus, one must summarize the potential use of oil in international disputes as having little more than nuisance value. Apparent short-term rigidities in oil supply arrangements, which indicate a possible means of exercising control and authority over unfriendly nations, quickly dissolve into relative insignificance given the alternative arrangements which can and will be made by the 'unfriendly' nation. In the medium term such rigidities tend to disappear altogether following the evolution of alternative strategies for obtaining oil requirements. Such 'oil crises' are therefore generally overplayed in the day-to-day reporting of events. Such reporting does not, in general, face up to the realities of the international opportunities for getting oil around the world, even to countries which most nations' statesmen would like to see boycotted.

With the exception of western concern over the U.S.S.R.'s oil diplomacy in the Middle East, it is significant that the post-war East–West struggle has not been over-concerned with oil as such. The Soviet Union has pursued a policy of self-sufficiency in oil, but has shown itself sufficiently flexible to arrange 'swop' and other deals with private enterprise companies in the West. On the other hand, N.A.T.O. laid down ground rules about trading in oil with the Soviet Union, but when these were ignored by Italy it did not appear to give rise to much real concern. It also attempted to restrict sales of certain oil industry equipment and pipelines to the U.S.S.R. and its allies, but the attractions of the Soviet markets for these types of exports from the industrialized countries made any such collective policy unworkable. First one ally then another made offers to the Soviet Union as a means of keeping surplus producing capacity in their steel and other industries fully occupied. Even in the revived 'cold-war' atmosphere of 1981 and with particularly difficult disputes over the number and deployment of missiles in Europe, it was significant that in the same year the Soviet Union successfully concluded contracts with a number of Western European countries for the sale of another 40 milliard m³ of natural gas per year from the mid-1980s. For their part the Western European nations concerned were anxious to participate in the arrangement, not simply as a means of achieving access to more gas, but also in response to the opportunity offered for supplying 5,500 km of large-diameter pipeline and a wide range of associated needs for industrial products in the construction of the system. The adverse reaction of the U.S. Reagan administration to this expansion of Soviet gas sales to Western Europe harks back to the earlier U.S. response to Russian oil sales in the late 1950s. It seems unlikely to be any more effective in the medium term though its importance should not be underestimated: U.S. opposition to increased West European imports of Russian gas has finally persuaded a number of governments in Western Europe that additional imports are probably not needed anyway – the gas required could be produced from a more active gas exploitation policy in Western Europe itself and it is this development which seems to be the most likely outcome of Reagan's policy. Indeed, the resolution of the dispute between the United States and its Western European allies over the Reagan administration's embargo on European oil firms' exports of goods involving U.S.-developed oil industry technology to the U.S.S.R. required the agreement of the European countries concerned to a U.S. proposal for an urgent study on the opportunities for increasing Western Europe's own gas production. The final results of this have emerged from the June 1986 agreement for West Germany, France and a number of other Western European countries to buy additional volumes of gas from Norway, from the giant Troll field which would otherwise have remained unexploited.

In other words, strategic concern over oil and gas trade tends to evaporate quickly because of the substantial economic gains that can be made by all parties concerned by keeping the oil and gas flowing. Such gains over the past thirty years of oil developments have included those accruing from the use of oil revenues by the major producing countries, the favourable economic effects of substituting high-cost coal with low-cost oil in countries like Japan and West Germany, and the general importance of the availability of oil products in almost all corners of the globe as a means of breaking the energy bottleneck in the development process. This well-nigh universal importance of oil in the process of economic development provided its most fundamental influence on the relationships between the world's nations up to 1973, and it gave rise to the whole complex set of issues that have been briefly examined in this book. However, as already indicated in previous chapters, the world oil system has undergone a quite remarkable shift in its organization and in its power structure since 1973, during which period the producing and exporting nations took control over the supply and price of oil by means of collective action. In so doing they threw the rest of the world into confusion over oil in particular, and over attitudes to energy in general, and thus created the need for a re-evaluation of the world of oil power. This important development is dealt with in Chapters 9 and 10, in as far as this was possible in late 1985, given the continued existence of a still very dynamic situation over the supply and price of international oil and the international relationships which emerge from this situation. A Postscript – written in May 1986 – to Chapter 10 attempts a last-minute update on the highly dynamic situation.

9. The Revolution in the World of Oil Power, 1970–75

It is more than seventeen years since the first edition of this book was written, in 1969. Experience throughout this period has effectively confirmed the point made in the Introduction that a day rarely passes without oil being in the news. Indeed, since 1969 oil affairs have become relatively much more important and thus the text of the original eight chapters has been greatly changed to take note of developments over this period. The world oil industry continued to expand until 1979 (except in 1975), albeit at a slower rate than in the twenty-five years from 1948 to 1973, so that companies in the industry continued to grow both in size and number, whilst the producing countries enjoyed rapidly rising government revenues from their sales of oil abroad and, in particular, from their successes in increasing their revenues per barrel of oil sold. Until 1979, throughout most of the world, oil (taken together with natural gas) continued to increase its share in the total market for energy.

One reflection of this trend was the success of oil in finally ousting coal from its position as the principal energy source in the United Kingdom in 1971. Britain's economy had been built up on the basis of the country's indigenous coal resources. As a result of continuing investment in the coal industry in the 1950s, a relatively low-paid mining force, and governmental help, the British coal industry managed better than in other countries of Western Europe to limit the competition from oil. Even so, in September 1971, oil in Britain took over as the most important source of energy in the economy. Thereafter, the use of coal continued to decline even after the severe oil price increases in 1973–4 and 1978–9 and, in the aftermath of the miners' strike of 1984–5, has since declined still further so that at least one-quarter of the industry that remains is at risk over the next few years. Meanwhile Britain's own production of oil and natural gas has expanded rapidly. By 1982 it provided over 55 per cent of the country's total energy production, with coal then responsible for under 40 per cent. In 1985 oil and gas's share had risen to two-thirds of the total.

In the meantime, on the other side of the globe, Japan leapt further ahead as the world's most important oil-importing nation and, in 1972, its imports exceeded 200 million tons for the first time. They then grew to over 225 million tons by the mid-1970s, making Japanese concern for its overdependence on oil produced abroad (and oil, moreover, brought to Japan almost entirely by non-Japanese companies) even more pronounced, as suggested in Chapter 6. This increasing concern has been reflected in Japan's widening and intensifying oil and gas operations. These are now proceeding apace, not only on the country's own continental shelf, but also, through the efforts of the state-aided and state-encouraged Japanese oil companies, in many countries in every continent of the world and especially in South East Asia and Australasia. Even in distant Western Europe increasing Japanese interest has been shown in the investment opportunities offered by the major oil and gas developments in the North Sea and on other parts of Europe's continental shelf. Indeed, indirect Japanese involvement in oil in this part of the world has now been initiated.

Elsewhere, the developing nations, under the impact of deliberately fostered industrialization policies, were generally unable to produce sufficient energy at home quickly enough to sustain their economic growth. Thus, they were obliged to import steadily increasing quantities of oil from the main oil-exporting areas. Though many of them tended towards the use of state trading and the construction of nationally owned refineries, etc., they nevertheless remained very dependent on the international oil companies for their increasingly important flows of imported oil.

At the same time the United States, as a result of domestic policies which constrained the rate of development of its indigenous oil and gas resources, rapidly increased its imports of oil from 1968 onwards and, as shown in Chapter 2, was thus obliged to revise and eventually, in 1972, to abandon its oil import quota system. This led to a sharp rise in the quantities of Middle Eastern oil which had to move to the United States. By 1978 almost half the oil used in the United States was coming from abroad, mainly from the Middle East. This necessitated the construction of crude oil terminals on Caribbean islands and in the Bahamas, where the mammoth tankers which have to be used to ship the crude oil from the Middle East in order to minimize transport costs discharge their cargoes ready for transhipment in smaller tankers to the United States. There, prior to 1982, no Gulf of Mexico nor east coast U.S. port could handle tankers of more than 100,000 tons.

More important, however, than the effect that this increasing use in the 1970s of internationally traded oil had upon shipping and infrastructure developments, was its effect upon the attitudes and policies of the different parties in the world oil power system. In order to appreciate the significance of this it seems appropriate to recapitulate the main elements of the system that had emerged between the late 1940s and the early 1970s in order to put the fundamentally changed situation of the present period clearly into perspective.[1]

The international oil system developed elements of instability shortly after the end of the Second World War. Until then, a few international oil companies, organized within the framework of a cartel which had first emerged to protect the industry's profits in the difficult economic circumstances of the 1930s, had largely controlled the supply and price of oil. There was little interference either from the countries in which production was concentrated or from the consuming countries, most of which used oil simply as a fuel to supplement the indigenous coal or hydro-electricity on which most of them depended. The international companies, moreover, effectively organized their activities around the world behind the guarantee of security offered by the political and/or military presences of the United States and the United Kingdom, which between them provided the home base for six and a half of the seven international oil corporations (with the remaining half, the Royal Dutch part of Shell, domiciled in the Netherlands).

Instability was introduced in this hitherto stable system when the United States government took action in the late 1940s to ensure that its tough domestic line against cartels and monopolies, enforced through the medium of its anti-trust legislation, was also applied to the foreign operations of American companies. This action, moreover, coincided more or less with another event which further undermined the traditional control of the oil industry by the oil companies. This was the insistence by the administrators of the Marshall Plan that the oil companies should cease to charge for the oil they sold to Western Europe (this oil was then largely paid for out of Marshall Plan funds), as though it were produced from high-cost fields in the United States and as though it were transported from the Gulf of Mexico. Thus, a new posted price system based on Caribbean and Middle East areas of oil production, and on freight rates related to actual, rather than hypothetical, movements of oil around the world, was introduced. It not only made the system more complex, but also introduced the prospect for the competitive marketing of international oil for the first time since the early 1930s.

At the same time there were other blows to the stability of the oil

1. The next few paragraphs essentially paraphrase the descriptions of development which have been treated at length in earlier parts of the book.

companies' controlled system which turned out to be even more significant in their effects. The most important was the rise of nationalism in the oil-producing countries. These countries, with their knowledge of the rapidly increasing demand for oil in an industrializing world, quickly came to appreciate that the exploitation of their oil reserves, in a physical sense, was also exploitation of their resources in an economic and political sense as well. Thus, the oil-producing countries took a series of steps to improve the rewards to them from the oil companies' operations. For example, in 1951 there was the initial agreement on a fifty-fifty sharing of the profits between Saudi Arabia and Aramco, and this was followed by similar agreements in most other oil-producing countries within a year or so. Later, there were other agreements between the two parties which brought changes in the concession-style arrangements, further tax rate changes and an early form of 'participation' by the producing countries in production activities. Venezuela, Indonesia and Iran played important innovative roles on these questions. Overall, the companies had to learn to live with increasingly expert opponents in the oil-producing and exporting countries, within the framework of a changing world political and economic environment in which they were viewed as the over-powerful neo-colonial exploiters of the resources of the oil-rich countries.

At the consuming end of the system, the industrial nations of Western Europe (and Japan) pursued policies which aimed at securing increasingly essential oil supplies (oil in most cases having quickly become the most important energy source for these countries), at as low a cost as possible. Some countries, especially Japan and France, controlled the international oil companies' activities very severely. Others, like Italy, sought to launch their own rival organizations to be responsible for national oil supplies. Still others, like Sweden and West Germany, encouraged the development of oil imports from countries where the production was not the responsibility of the international oil companies, notably from the Soviet Union – which made a post-war reappearance on the oil market in the late 1950s – in order to bring downward pressure to bear on prices, and to achieve a higher level of security in respect of their essential energy supplies. Ironically, however, it was again the United States itself which was mainly responsible for undermining the prices at which oil was delivered to importing nations in Western Europe and elsewhere. This occurred when the United States determined in 1959 to pursue a protectionist oil policy and to limit its imports to about 12 per cent of the expected demand for oil in any one year. Over the previous few years a large number of American domestic oil companies (as distinct from the American-based international companies) had established operations abroad for the first time in order to find lower-cost oil which they could then profitably market in the U.S.A., where costs of domestic oil were rising quickly. Their anticipated sales to the United States were, however, blocked by means of the newly introduced U.S. oil quota import system and they had to seek markets for their newly developed oil production in other parts of the world. In these markets the combined impact of additional supplies and of new companies aggressively seeking marketing outlets (in contrast with the international majors which preferred 'orderly' marketing) brought a cheap oil era to Western Europe, Japan and elsewhere and a steady decline in the overall profitability of most of the oil companies.

Thus in a situation of pressure, on the one hand, from the oil-producing countries for more tax revenues, and price weakness for most oil products in many of the world's markets on the other, the oil companies became something less than the high profit centres they had been in the early post-war period. This situation, coupled with the complicating and adverse effects of political difficulties in the Middle East, from which area a rapidly increasing share and a growing volume of the crude oil supplying the world markets originated, made the outlook less than favourable for the companies. It set them scurrying into diversification exercises and into a search for a solution to the instability of the world oil market.

The evolution of this pattern of development had, of course, led to the general ability of the world's oil consumers to secure their increasing needs at a steadily falling real unit cost. As Figure 1 (p. 142) shows, the price of oil fell in real terms more or less continuously after 1950. By 1970 it was only 40 per cent of the price twenty years previously. This price decline also produced a situation in which other energy sources had found it increasingly difficult to compete with oil. The combination of these factors led to the very high growth rates in oil demand which were experienced over the period. For the oil-producing countries, the international oil companies and the U.S.A., however, the situation by 1970 had for different reasons become unacceptable. All three parties were thus both motivated and anxious to see the situation changed and they initiated steps to secure their objectives.

The initial requirement for a fundamental change in the oil system was a re-evaluation by the oil-producing countries of their status and strategy. This, in turn, required much more effective political action than they had hitherto been capable of taking. Such action on the part of the oil-producing and exporting countries did, however, become possible in the late 1960s and when it was taken it initiated the important new developments which have affected oil and world power in the last ten years. The overthrow of King Idris of Libya led to the establishment there of the revolutionary government under Colonel Qaddafi without any attempt by Britain and the United States, both of which had thousands of troops in Libya, to prevent the change from the system of government which they had installed and with a king who remained favourable to western interests. The result was the strongest political action by a single oil-producing country against the oil companies to which it had granted concessions since Mexico's nationalization of its oil in 1938. The new Libyan government was, however, in a position to get tough with the companies concerned for several reasons: first, it had the strength which was created by the large revenues from the rapidly-rising oil production of the large number of corporate groups which had been granted concessions to operate in Libya since 1960; second, there was the favourable impact of events in the Middle East on the country's competitive position in the world market; and third, there was the failure of Britain and the U.S.A. to intervene to stop Qaddafi from achieving power and then from exercising it in a way which seriously affected the interests of the oil companies. This he did by using appropriate tactics. He tackled the companies one by one and threatened each of them with the forcible curtailment of their planned levels of production. In the favourable market conditions of the time for oil from Libya this was generally enough to persuade the companies to agree to raise the price at which they 'posted' their oil for export, and thus automatically pay increased revenues per barrel to the Libyan government. The few companies that chose to test the revolutionary government's intentions in this respect quickly found that it meant what it said and so they fell into line – or else they were nationalized. Libya, in effect, was beginning to exercise control over the supply and price of its oil exports.

Libya's get-tough policy was quickly noted by other producing countries. Some of them took similar action and also secured agreements by the companies to pay increased revenues to the countries concerned. Hitherto in the oil world unilateral action on the part of the individual countries had, as shown in earlier chapters, been constrained by the fear that the companies concerned would cut back their oil production and investment plans in the country and then make it up by expanding their output and activities elsewhere in the better-behaved parts of the oil-producing world! It had, of course, been partly in response to this fear that O.P.E.C. had been formed. Now it, too, given the changed attitudes of its member countries, together with the impact of significant changes in the post-1970 overall international oil supply/demand relationships (see below, p. 225), was able to take collective action of a kind which had previously been impossible, because it would have been rendered ineffective by competitive (rather than cooperative) responses by its member countries.

The success of O.P.E.C. in achieving a consensus amongst its members for collective action – as had been agreed at its Caracas meeting in December 1970 – marked the beginning of a significant change in the power balance between the various groups with an interest in oil. O.P.E.C. now became a sort of cartel of the oil-producing countries. Its initial success, however, was not unrelated to the fact that it was also accepted and even welcomed in this role by the same oil companies which, when it was created as an organization ten years previously in 1960, had declined even to recognize its existence at all. Later they had come to view it as something of a nuisance, but certainly not as an organization to be taken seriously. This success of the oil-producing countries in forming a politically and economically significant organization, in spite of the way it had been treated, had, of course, significance even outside the framework of the oil industry, for O.P.E.C.'s success in creating an inter-government cartel was the first amongst producers of primary commodities at an international level. Thus, it seemed likely at one stage to provide a model for the producers of other commodities, such as bauxite and copper, which were also not only essential to the economies of the world's industrial nations, but whose large-scale production was restricted to a relatively small number of developing countries.

To return, however, to oil itself. Events between the late 1960s and the early 1970s also produced another equally significant change in the international oil industry's organization. This was the move that was made towards the re-establishment of an international oil agreement by the companies responsible for most of the oil that entered world trade. Their motivation was clearly the protection of their already deteriorating level of profits in a situation in which they expected to have to continue to pay steadily increasing taxes to the producing countries, and thus needed to be able to pass these increased costs on to their customers. Since the 1940s, the main barrier to collective action on the part of the companies had, as shown above, been the United States' legal restraints on such agreements through the working of internationally applicable anti-trust legislation developed, in part, because of earlier oil industry propensities to act collectively! The oil companies, however, in the period immediately after the Libyan and O.P.E.C. action described above, managed to persuade the Nixon administration that there was a real danger of a crisis in the world oil industry and that the best way of preventing the crisis lay in the companies being permitted to take concerted action in their negotiations with the producing governments. Thus, early in 1971, the companies obtained permission from the U.S. government to work together to this end and so were able to reach collective agreements with the oil producers. This was a development for which the companies' strategic policy-makers had striven since 1968 when the so-called London Oil Policy Group, formed by the main international oil companies in the aftermath of the 1967 Suez crisis, met together for the first time to work out a strategy for bringing order back to the world oil market. Though the companies publicly maintained the appearance of fighting O.P.E.C. tooth and nail over the agreements on prices and taxes demanded from them – as they sought to keep their increased tax obligations to a minimum and to delay 'participation' in (that is, the nationalization of) their oilfields and other assets as long as possible – they certainly recognized that their best hopes of future profitability, and even of survival, depended on their achieving a *modus operandi* with the major oil-producing countries. The latter, in turn, quite readily accepted the idea that they needed to work with the international oil companies in order to ensure that their oil, and thus their revenues, continued to flow. Thus O.P.E.C./oil companies cooperation became a fact of the oil power system of the early 1970s, with the positive encouragement of the United States.

The U.S.A. for its part wished to see the establishment of a new collective stability in the oil system for two reasons. In the first place, it sought to establish a basis for a renewed effort to find a political solution to the Middle East conflict. In this respect, it argued that higher revenues and a greater degree of economic certainty for the Arab oil-producing nations would make it easier for them to accept a compromise in their dispute with Israel, and so bring greater political stability to the whole of the Middle East.

Additionally, the U.S.A. was fed up with a situation in which the rest of the industrialized world had access to cheap energy (which the U.S.A. itself could not have because of its policy of protecting indigenous energy production, both oil and coal). Thus, it deliberately initiated a diplomatic effort which aimed at getting oil-producing nations' revenues moving strongly up. It did this by having its representatives talk incessantly to the oil-producing countries about their low oil prices and by showing them the favourable impact of much higher prices. It could, of course, rely on the cooperation of the largely American oil companies for ensuring that any consequential price increases which the O.P.E.C. countries introduced, plus further increases designed to ensure higher profit levels for the companies, were passed on to the European and Japanese oil importers and so raised the cost of the energy used. In this way the energy cost advantages of their industries over their competitors in the United States would be eliminated. In as far as the U.S.A. itself would be affected by the higher foreign exchange costs of the increased amount of foreign crude oil that it expected to have to import, this, it was argued, would be offset entirely or to a large degree, by the greatly enhanced abilities of the U.S. oil companies to increase their earnings and thus their remittances of profits back to America.

Within the framework of a re-evaluation of how their best interests could be served and the consequential establishment of a somewhat 'unholy alliance' between the United States, the international oil companies and the O.P.E.C. countries, the stage was set for changing the international oil power situation that had evolved over the previous two decades.

From the companies' point of view, of course, their acceptance of the producing countries' demands for increased taxes and the idea of participation by the producing countries in the production enterprises (which was, in essence, another device for increasing the share of the profits going to the countries), necessarily meant a major increase in their tax-paid costs. This meant that the weakness in the international oil market had to be eliminated if the companies were to earn what they considered to be adequate profits. Thus, the actual timing of O.P.E.C.'s success and the companies' acceptance of it can be correlated with the strengthening of the international oil market after 1970. This was due to the combination of a set of unusual circumstances: first, a strong demand for most oil products in most markets, in a period of general economic advance; second, a shortage of refinery capacity in Europe and Japan so that the demand for oil products exceeded the supply; and third, a temporary scarcity of tankers which was aggravated by the politically occasioned closure of the Trans-Arabian pipeline. The fact that oil prices strengthened as a result of these circumstances gave the oil companies the public relations opportunity to persuade oil consumers that there was an oil supply crisis. This they did, not only over the short-term prospects for oil, but also in respect of the long-term oil supply/demand outlook which, for the first time in the post-1945 period, was presented as one of potential scarcity. They argued further that these facts, coupled with the imposition of the swingeing new taxes in the producing countries, inevitably meant significant and continuing rises of oil prices over the whole of the foreseeable future. The U.S. energy crisis at the time added further strength to arguments about the inevitability of higher prices for oil, even though the crisis there was an essentially short-term one, arising out of domestic energy pricing issues, and from the impact on the development of energy supplies of the conservationists and from the environmentalists. The public, and governments, were persuaded of the oil companies' case, not least because of the coincidental publication in 1972 of the polemic, *The Limits to Growth*, in which the inevitability of global raw material scarcity, including energy, had been argued.

A climate, in other words, had been established for an attack on oil consumers' interests in which the aim was the elimination of the cheap energy (and essentially low-profit energy) to which Europe, Japan and many countries elsewhere had, by 1970, become very accustomed. Temporarily, it was thought, the factors in the short-term oil supply/demand situation mentioned above could be relied on to secure higher prices for the initial period, but their essentially ephemeral nature meant that the safeguarding of the longer-term position required positive action on the part of the companies. Fortunately, their new-found ability and opportunity to work together with the oil-producing countries gave ample opportunities for strategic marketing discussions, and for decisions on how best to tackle the inbuilt propensity for price weakness in Europe.

This market was the critical one to bring under control for it was to Europe in the past ten years that 'distress' supplies had found their way in their search for outlets and so generated weak price conditions. The appropriate device to achieve this end was obviously a moratorium on the expansion of the infrastructure through which oil was moved and marketed on the continent (particularly in refining and pipelining facilities), accompanied by some element of rationalization in the complex company structure involved in the marketing of oil products in Europe. Thus, expansion projects were slowed down and some companies -- for example, Gulf, Shell and B.P. - decided to pull out of certain national markets and/or out of the sale of specific products. Meanwhile, in the case of some products - for example, aviation fuels - there was evidence for an agreement whereby existing suppliers of specific airlines and airports were accorded the right to continue to have the business as other companies declined to bid for it or else deliberately bid higher prices than the existing supplier. All this, of course, fell far short of the establishment of a formal cartel of oil companies, or even a repetition of the 1933 'As Is' agreement, under which the oil companies had formally agreed to leave market shares as they were. However, given the high degree of understanding between the companies, and the fact that almost all of them were involved in the collective discussions with the oil-exporting countries, so that they were aware of the essential facts of the situation, there seemed to be good prospects of agreement on ways of getting a higher degree of 'sanity' and of 'orderly marketing' into the hitherto cut-throat and generally chaotic oil market in Western Europe.

At first, however, unforeseen factors combined to undermine the strategy (though it should, nevertheless, be noted, as can be seen in Figure 1 (p. 142), that most oil prices in most countries did go up after 1970 by more than enough to compensate the companies for the higher taxes they were having to pay in the producing countries, and for other increased costs, with, of course, a favourable effect on oil company profit margins). In the first place the weather was against the oil companies, for Europe had a series of warmer than average winters and this played havoc with the expected demand for heating oils. Likewise, demand for industrial fuels was lower than expected because of a slowdown in the rate of European economic growth. The companies recognized the temporary nature of these two factors and were thus prepared to sit out their influence.

The third factor, however, was potentially much more dangerous for the oil companies' strategy in that it threatened to get worse over time. This was Western Europe's newly found large-scale sources of natural gas which, as the fuel preferred by most customers in a wide variety of end-uses, threatened to lead to stagnation in the growth of the markets for fuel and heating oils. This factor, however, was eminently controllable in that most of the gas production in Europe was at that time the responsibility of one or other of the international oil companies. These thus had the option, on the basis of one pretext or another, of ensuring that natural gas production was held back. This happened in both the Netherlands and the United Kingdom, where the development plans for the expansion of the new energy source were significantly restrained. With the danger of supply expansion out of the way, it could thus be assumed that the worst effects of an expansion of the availability of natural gas in Western Europe on the profitability of oil markets would be over by 1974. Then, given average or worse than average winters, plus a resumption of industrial growth on the continent, the increased demand for oil could be expected to lead to much higher prices for most oil products. And with the European market under control, as orderly marketing replaced the competitive situation of the twenty years since the mid-1950s, the oil companies could then quite reasonably assume that markets elsewhere in the world would take care of themselves. Increased taxes and cost increases in oil production, the results of changing relationships with the member countries of O.P.E.C., could then be more than passed on to consumers, with consequently highly favourable conditions for the companies to earn higher profits.

In brief, the competitive nature of the oil market between the middle 1950s and the early 1970s seems to have turned out to be an aberration. Since then, there has been a reversion, as a result of the factors outlined above, to the more usual oil industry pattern of producers' control or attempted control over the supply of oil. The new strategy of producers' control was fully intended to be under the leadership and direction of the major international oil companies; even though they, in turn, certainly recognized that the greater part of the enhanced profits to be made out of the policy of restraining the supply of oil would flow to the producing countries, whose interests, the companies thought, would thus be fully and satisfactorily served. Thus, the Teheran and Tripoli and other negotiations in 1971 and 1972 between the international oil companies and the oil-exporting countries should be interpreted as an attempt by producers – both companies and governments – to achieve a satisfactory *modus vivendi*, within the framework of which somewhat increased profits and greatly increased revenues, to the companies and the exporting countries respectively, could be achieved. What the companies had in mind was the establishment of more profitable orderly oil marketing in place of the highly competitive market situation with its limited profitability of the previous fifteen years.

The success of such a strategy depended not only on the producing countries' willingness and ability to work together – a development which was achievable and, indeed, achieved through the increasing effectiveness of O.P.E.C. – but also on their willingness to continue to accept the idea that the major oil companies took the leading role in the organization of the international oil industry. This implied not only the continuity of companies' responsibilities in respect of transporting, refining and marketing the oil, but also their role as decision-takers on such fundamentally important matters as levels of production and the continued development of producing capacity in different countries. In the latter respect the companies saw their responsibility as one in which they continued, on a long-term basis, to expand the supply of oil in line with their expectation that its use would continue to grow at its historic average of 7 to 8 per cent per annum for another fifteen years or so, and thereafter at a rate of 5 per cent per annum for the rest of the century. In 1970 the non-communist world used over 2,000 million tons of oil. By the year 2000 the oil companies then anticipated, the industry would need to supply over 8,000 million tons and the companies were confident that they would be able to achieve the continued expansion of the global oil industry that such an increase in demand over the next thirty years required.

This grand strategy for a post-1970 orderly, though still rapidly expanding, world of oil presented the prospect of a very significant change from the oil power situation of the previous fifteen years. Even this radical prospect was, however, soon undermined when the oil-exporting countries realized that they did not actually need to cooperate with the multinational oil companies. This happened when they recognized, in essence, that they could themselves take over from the companies the latter's control over questions of price and of the levels of production which were to be established. Even during 1971 and 1972, when the negotiations on the 'grand strategy' between companies and exporting countries were still being actively undertaken, there were already many straws in the wind which indicated that the oil-producing nations were moving towards the assertion of control over supply. These straws came in various forms. They included the expropriation of company assets (as, for example, in Libya and Algeria); unilateral decisions to fix maximum rates of production (in, for example, Kuwait); the unwillingness of other producers to accept the production expansions which had been scheduled by the companies (for example, by Saudi Arabia); and increasingly close national attention to, and concern for, the horse-trading between companies of their oilfield assets in producing countries (as, for example, by Abu Dhabi). By early 1971 it was already possible to see the imminent development of an oil-supply and pricing crisis.[2] This arose from the eminently reasonable expectation at that time that producing-country control over production and development decisions would gradually become the norm. Such unilateral decision-taking by the oil-exporting countries would thus replace the apparently agreed bilateralism between countries and companies, in which the latter would have restrained the rate of change of the price of oil in the interests of maintaining the rate of growth in the use of oil for which they had already made medium- to long-term investment plans.

The renewed outbreak of war between Israel and the Arab states in October 1973 and the latter's ability to use their dominance of the oil

2. The author pointed this out as early as February 1971 in an article, 'Against an Oil Cartel', in *New Society*, No. 437, 11 February 1971. Unhappily his warning of impending dangers was denied by the oil industry (in a response by the head of Shell's Government Relations Division in *New Society* later in February 1971) and was ignored in the oil-importing countries where the governments blithely assumed that the oil companies were protecting the interests of oil users and that the companies would keep cheap oil flowing in the increasing volumes required.

market, and the dominance of oil in the western world's energy market, as an economic and political element in their struggle, created the opportunity for the accentuation and acceleration of the process of establishing producing-countries' control over the supply and price of oil. This opportunity arose from the fact that two of the major industrialized regions of the world – Western Europe and Japan – had not only allowed themselves to become almost entirely dependent on O.P.E.C. oil for sustaining their economic systems, but had also failed to take any effective counteraction to protect their oil supplies and to minimize their use of oil, even in the context of a clearly deteriorating outlook for the viability of their cheap energy policies in the changed post-1970 oil situation. Western Europe had also failed to take seriously the option of a more autarkic energy policy, the possibilities for which had been opened up by the prospects for North Sea oil and gas production on a very large scale indeed.

Within three months of the outbreak of the 1973 Arab–Israeli war there emerged a state of imbalance between likely available oil supply and the potential oil demand which would, without the war and the opportunity it gave to control supplies, have taken perhaps three years to develop. Had these three years been available and used to good effect then there *might* have been time enough for adjustments to have been made to the structure of demand and to the geography of energy supply – though, as pointed out above, there was no evidence between 1971 and 1973 of any European or Japanese realization that radical action in respect of energy was in fact required. By 1973 the required adjustment to the new international oil supply situation became harsher and much more difficult to make, without causing serious problems of unemployment and a supply-generated depression in the western world. It demanded, in effect, a range of decisions whereby life-styles could be changed to a marked and, perhaps, a hardly acceptable extent.

The chances for action even in 1973 were, however, still thwarted by the absence of any general acceptance – either in Europe and Japan, or in the United States – that the world of oil had undergone a near-instant revolution rather than merely a radical change, and that there was unlikely to be a return to an unrestricted availability of supplies of O.P.E.C. oil as hitherto in the days of oil companies' control over production levels, even if the politics of the Arab–Israeli dispute were settled. The importance of the unrecognized changes in the world of oil power which had taken place can, however, be seen in the absolute control which the producing nations took over decisions on the level of posted prices, on the amount of oil to be produced and on their ability to choose those customers with which to trade. As far as the Arab oil-producing states were concerned there was, of course, no doubt that these decisions implied the use of oil as an economic weapon in the struggle with Israel. The fact that there was a much more fundamental change of attitude at work amongst the oil-producing countries was, however, clearly expressed in decisions on oil supply and price taken by the five non-Arab members of O.P.E.C. Thus, it was Iran which started the procedure of auctioning its royalty oil (rather than selling it to the companies at a price related to the posted price) and, having secured up to $17·30 per barrel for the oil it auctioned (compared with a posted price for the oil involved of $5 per barrel), Iran then led the December 1973 move for a further doubling of the officially posted prices in order to relate oil prices more closely to what consumers were apparently prepared to pay on the open market. Iran, moreover, still talked at that time in terms of the need for, and the validity of, yet higher prices in the near future.

Ecuador, the newest O.P.E.C. member and a potentially important alternative source to Middle Eastern oil for the United States, unilaterally increased its already high posted price to a level higher than elsewhere in the world to take advantage of its special position *vis-à-vis* the United States. Meanwhile, Nigeria, a member of the Commonwealth and presumably, therefore, an even closer friend of Britain than the Arab states were thought to be, declined to increase its oil exports to the U.K. and took full advantage of the constrained supply to maximize its revenues. And, finally, the first act in January 1974 of the newly-elected President of Venezuela was to announce that his country had no interest in increasing oil production beyond the levels already reached. He later indicated that reduced levels of production would be more appropriate in order to ensure the continuation of price increases – the type of approach to the oil price question which Venezuela had propagated for ten years with its fellow O.P.E.C. members before finally getting them to accept the strategy.

For non-Arab as well as Arab oil exporters, their motivations and ability to keep the oil supply constrained thus became increasingly strong. This arose, first, from the way in which the exporting countries undermined the power of the multinational oil companies and turned them instead simply into the agents for implementing the essential decisions over the supply and price of oil which the countries concerned had already taken; and, second, as a result of the degree to which the exporting countries' efforts massively to increase their return on each barrel of the limited supply of oil they made available had been so immediately successful. This success had, moreover, then been maintained against the hostility of the industrialized countries.

As a result of these revolutionary forces O.P.E.C. oil became very high-cost energy indeed for the importing countries. Behind this incontrovertible fact, however, there lay at the time strongly contrasting interpretations of the strength and motivations of the oil-producing countries, and hence contrasting ideas about the strategies that ought to be followed in response to the very serious situation that had arisen for the western world's economy.

Many industrialized countries were relatively quickly persuaded that the oil exporters were both willing and capable of maintaining their policy of supply limitation and increasing prices. Thus, they argued, strategy should be based on a package of measures designed to ensure that O.P.E.C. cooperated in getting the oil moving in the volumes which could be calculated to be required for the next five to seven years at prices which the rich, industrialized, importing countries, even if not the Third World importers, could just about afford! This then required the further optimistic assumption that the oil-producing lands could be persuaded to continue to circulate in the international monetary system the enormous oil revenues which they were expected to receive (over $60 milliard in 1974 and at least $100 milliard in 1975). This would provide both the mechanism for automatic adjustments of balance of trade difficulties, arising, in part, out of higher oil prices, and the means whereby demand in western economies could be kept at a high and expanding level, so enabling international economic development to go ahead. The strategy further assumed that western world inflation could be controlled and moderated, in spite of the unfavourable impact of continued rises in the price of oil. It also conveniently glossed over the high probability that something would go wrong, albeit only accidentally, in the proposed untried and fragile system which would be subject to all kinds of economic and political pressures. The optimism appeared to stem from the lack of appreciation of the revolutionary change that had occurred in the world of oil power, and the failure or the unwillingness to recognize that, for the first time in some 400 years, the western world had lost control over an essential element in its system to a set of countries which had hitherto not been considered as decision-taking entities of any international significance within the system.

A strongly contrasting view of the situation emerged from the recognition of this revolutionary change in the oil world. From this it was argued that the future of the world economy depended on the policies and actions of a group of countries which had not hitherto had such power and responsibility. It was, moreover, impossible to be optimistic about the likelihood of these countries using this power and responsibility in order to enable the industrialized countries of the western economic system to continue to enjoy their privileged position in the world.

The lack of optimism arose from three considerations. First, the new 'masters' lacked the experience and background adequately to gauge the results of their policies: in other words they could accidentally destroy the fabric of the system as a result of their inability to judge the impact of their actions. Second, some of them did not necessarily want the western economic system to survive, particularly bearing in mind the 400 years during which they had been the victims of western economic exploitation. The undermining of the system could thus be considered a prerequisite for a more appropriate world order. Third, as most of the O.P.E.C. countries gave high priority in their policy-making to the elimination of, or to their control over, Israel, this brought them into confrontation with the western world which, to the Arabs, was the mechanism whereby Israel had been created and continued to exist.

In light of these three factors the chances of the oil-exporting countries moderating and tempering their demands sufficiently to enable the western economic system to survive appeared to be small, and they grew smaller as time went by without any agreed western response to the situation. As a consequence there developed a threefold threat not only to the future economic viability of the west, but also to the very existence of the sort of society which had emerged in the western world.

The threat consisted of the following elements. First, there was the basic question of the availability and security of oil supplies from O.P.E.C. countries in a situation in which almost all the rest of the western economic system (including the Third World oil-importing countries) had come to depend on a steadily increasing flow of imported oil to sustain the rising standards of living and the accepted ways of life of its inhabitants. The likelihood that there would be interruptions in these supplies from time to time, and the high probability that there would be a decision by the producing nations permanently to cut back the amount of oil exported to the west, served to undermine the planning and the policies of all western governments and so make it impossible for them to sustain the rising expectations of their populations for continued development.

The second aspect of the threat emerged from the hyper-inflation generated by the traumatic increases in oil prices. In this respect there was a widely-accepted view that the increases in oil prices would simply create a once-and-for-all problem of readjustment to economies based on more expensive energy. These hopes, however, depended on a decision by the oil-exporting countries not to index their oil prices to the inflation rate in the western world. Such indexation was, in effect, more than maintained during the rest of the 1970s and O.P.E.C. was committed to use it as the basis of its long-term pricing policy. This, of course, had the effect of further strengthening the initial serious impact of increases in oil prices on the inflationary trends in the world economy and of creating a danger of the whole fiscal system going out of control, especially in a situation in which so many internal costs and prices in many western industrial countries are also indexed to inflation. O.P.E.C.'s pricing policies thus opened up the real possibility of internally and externally generated cost and price increases in the industrialized countries feeding on and sustaining each other.

The collapse of the western system also became possible from a third threat created by the changed world of oil power structure. This was the difficulty arising from the tremendous surpluses on their foreign trade accounts which the oil-producing countries began to build up from their sales of high-price oil. In 1972 these countries earned less than $5,000 million on their sales of oil abroad, and spent most of this on buying goods and services from the west, so keeping the money flowing round the system. In 1974 their revenues surplus to their immediate import needs were over $30,000 million and thereafter they generally continued to increase each year through the rest of the decade, under the impact of rising oil prices and continued relatively high levels of exports of oil from the O.P.E.C. countries. In such circumstances the oil-exporting countries were able to choose their monetary strategy. On the one hand, they could choose to do nothing with the money (which, indeed, had little value to many of the countries concerned) and, by withdrawing so much purchasing power from the international economic system, intensify the world recession. Or, on the other hand, they could choose to be positively unhelpful by, for example, using their monetary reserves to manipulate money markets, or even individual institutions through their use of very short-term loans. On the basis of this sort of behaviour it becomes very difficult to recycle the money in any economically useful way without exposing the lending institutions to high risks. Through the use of the many unhelpful ways open to them, the O.P.E.C. countries had the ability to bring the world's monetary system under severe pressure.

The combination of the impact of all these factors had already seriously affected the western world's economy by 1975, by which date there could no longer be any doubt over the new importance that oil power had secured in the world system. The industrial countries to some degree, and most non-oil-exporting Third World countries to an even greater degree, were by then suffering from the consequences of the quite fundamental changes in the international oil system, and a realistic interpretation of the prospects at that time were for a situation which would become progressively more serious. In brief, and quite simply, power in the system had been taken over by the oil-producing and exporting nations, working together through O.P.E.C., and as, in the short term at least, they seemed likely to act irresponsibly, the fundamental question thus became whether the international western economic system could be expected to survive. Though there was much that was and is wrong and inadequate about the system, and about individual parts of it such as the international oil component, its demise would not automatically serve to produce anything better, and most certainly not in the short term. It was in the context of such an extreme possibility that a view had thus to be taken of attitudes and policies to be adopted towards O.P.E.C. and its member countries by the hitherto unchallenged decision-takers – the United States and the other industrialized countries – in the western system. The evolution of these attitudes and policies since 1975 is described and analysed in Chapter 10.

10. The Response to O.P.E.C.'s Challenge

By 1975 the United States, conscious of the imminent likelihood of the enormous strains on the western system as a result of the revolution in the world of oil power over the previous five years, had had to reverse its earlier policy of seeking cooperation with the O.P.E.C. countries as a means of securing a more competitive energy situation for its own consumers and as a device for making the Middle East more secure (see above, pp. 224–5). Instead it began to see confrontation as the required, indeed the only, alternative. The formation of the International Energy Agency (I.E.A.) by the United States was related to this policy aim. It was constituted as a means whereby the world's rich oil-importing nations' strategies and tactics towards O.P.E.C. and the challenge of high prices for, and politically restrained supplies of, oil could be coordinated and strengthened. Even the stronger option of a military solution was examined and appears to have been rejected mainly because it would have required unacceptable collusion with the Soviet Union. Such collusion was not an impossible development, given the fact that economic development in the U.S.S.R. depended, then and now, on the continuation of a prosperous western world. Thus, the Soviet Union's analysts might well have suggested that, as such a prosperous western world was becoming impossible with the continuing oil crisis, the idea of the overthrow of O.P.E.C. and/or individual members of the organization ought not to be opposed. From the west's point of view, however, the military solution would have undoubtedly brought in its wake more serious problems than those it solved, as it would also have led to a generalization of the Third World countries' disillusion with the western system, and so closed the ranks of such nations in support of O.P.E.C.

Other countries presented other ways of dealing with the situation. France saw the required strategy as one which emerged simply from the acceptance of the irreversibility of what had happened. If one accepted the inevitable, France argued, one could then aim to secure a calculated required flow of oil, the quantity of which would be guaranteed (thus making national planning much easier), and the price of which would reflect the terms that could then be obtained, rather than the still higher prices which, France argued, the O.P.E.C. countries would soon undoubtedly seek and probably obtain. In return for the required flow of oil, there would be guaranteed outlets in the O.P.E.C. countries for advanced technological hardware, and associated managerial and other services. Furthermore, on the basis of such a mutuality of interests it would be possible to develop the necessary *rapprochement* with the relatively unsophisticated O.P.E.C. countries, and so win them over to help sustain the system through their effective cooperation with the industrialized countries.

It must be remembered, however, that 'special arrangements' in the oil world were by no means new and, in particular, that France had previously pursued such a policy following the independence of Algeria, for a combination of political and economic reasons. The background to this French approach in the 1960s and the results achieved thus constitute a valid precedent against which to evaluate the French plan for dealing with O.P.E.C. in the 1970s. This is not only because of the international implications arising from the search for special relations, but also because of the implications that any oil eventually imported as a result of the establishment of special relations has to be found a market, thus generating a requirement for a government-controlled refining and marketing policy.

The historical reasons for the development of French policy towards the international oil industry lie beyond the scope of this chapter, except to point out, as shown in Chapter 8, that it emerged out of the disconcerting effect, in the 1930s and the early post-war years, of French exclusion from an interest in the important oil-producing areas of the Middle East – a region of longstanding concern to France – as a result of the very powerfully backed diplomacy of Britain and the United States. In French eyes, the international oil industry developed into something which was little short of an Anglo-American conspiracy.

France's initial search for independence from this 'conspiracy' lay in a state-sponsored and state-financed exploration in overseas parts of the Union, most notably Algeria. Ironically, really significant success in France's search for oil pre-dated Algerian independence by only a few years, so frustrating French efforts to achieve independence in oil.[1] The negotiation of Algeria's independence was, of course, partly concerned

---

1. Ironically, the main benefits of France's large expenditure on oil exploration in North Africa probably flowed to the Anglo-American companies, given the incentive provided by Algerian discoveries for them to move quickly into neighbouring Libya to confirm that the North African oil province extended out of the area of French influence. Libya soon proved to be a more prolific oil-producing country!

with the need to reach an agreement acceptable to France over Algeria's newly discovered oil resources. France's success in eventually negotiating a privileged position with respect to oil exploration and exploitation in Algeria was initially interpreted as an indication of her ability to secure a special relationship with an oil-producing country where Anglo–American oil companies could not. With hindsight, however, one can see that it was really Algeria rather than France which needed such a relationship at that time. Algeria immediately after independence was, in fact, neither politically nor economically attractive to the international oil companies. At that time, the companies had access to more oil than they knew what to do with, as a result of a decade of great exploratory successes in the 1950s, from countries which offered greater stability and security than Algeria. Thus, any interest they might have had in Algeria would have given that country's oil potential low priority indeed for development, as compared with better opportunities in many other parts of the world. Algeria thus gained from the special relationship with France as it was able to maintain its oil as 'franc-zone oil' and so secure guaranteed entry to the French market. There the system of controls over refining and marketing in general, and the controls over the activities of the international companies in particular, meant that there were opportunities for selling Algerian oil at prices well above the going international market rate. Moreover, as the international companies operating in France were obliged by their agreements with the French government to use a certain percentage of Algerian oil, an incentive was created for them to invest in Algerian oil resources.

Thus, the special relationship between France and Algeria was achieved in part at the expense of the international oil companies. As a result of it they certainly lost opportunities they would otherwise have had for moving larger quantities of their own lower-cost crudes from other countries of the world to France, and thus they lost the profits which they could have earned on such oil supply arrangements. This loss of profits, however, was offset in large part by the fact that the generous refining and marketing profit margins built into the French-controlled system at the time enabled them to make higher profits in France on all the oil they sold there than they would have made in a free market using their own crude oils. Intensive competition elsewhere in Western Europe between an increasing number of suppliers had, by that time, brought the price levels of most products down to lower levels than the guaranteed prices which Algeria was able to secure from its sales to France and which, in turn, formed the basis for oil companies' prices in that country.

Thus the success of the special relationship between France and Algeria depended on France's and French consumers' willingness to pay more for their oil supplies than would have been necessary in the absence of the relationship. This willingness was derived, apart from the 'anti-conspiracy' consideration, from the search for the security of oil supply which the link between France and Algeria was supposed to produce. Even this, however, turned out to be somewhat illusory. Once the economics and politics of the international oil system changed, Algeria proved to have little or no compunction, first, in unilaterally changing the terms of the special agreement – by making the oil even more expensive to France – and then in revoking it. It not only demanded a price for its oil in excess of that which even France was prepared to pay, but also required the withdrawal from Algeria of the French oil companies, through whose activities the physical flow of oil had largely been maintained, not only from Algeria to France, but to other parts of the world as well. Thus, in spite of the special relationships, the French oil companies in Algeria managed to survive only a little longer than the non-French companies which had either withdrawn or been expelled in the previous two or three years. French or non-French essentially made little difference in a situation in which Algeria wanted to go it alone and felt that world oil politics and economics had so developed as to make this possible.

In the meantime the idea of special relationships as the basis for its policy towards oil had been further extended by France. Politically this followed from the country's *rapprochement* with the Arab world after the Middle East war in 1967. (Alternatively, the oil policy can, of course, be viewed as an inherent part of the general agreement with the Arab nations or, even, as one of the main motivations for seeking the *rapprochement*, given that difficulties with Algeria were already becoming apparent by 1967.) In practical terms the special relationship policy required the creation of a strong, entirely state-owned French oil company. From this policy there emerged E.R.A.P. (a state company based initially on the merging of pre-existing smaller state oil entities), which was to accept the responsibility of promoting French interests in as many oil-producing, or potentially oil-producing, countries as could be persuaded of the validity of a special relationship with France. This persuasion was not unsuccessful, and E.R.A.P. quickly became a concessionaire (or a contractor on behalf of a local oil state company) in Iran, Iraq, Saudi Arabia, Libya and elsewhere, as the French aim of having French companies produce an amount of oil at least equal to the total consumption of oil in France was vigorously pursued. The idea of a special relationship emerged in part out of France's arguments that its exploitation of another country's resources was qualitatively different from that of the Anglo-Americans. It was due in much larger part, however, to E.R.A.P.'s general willingness to pay more than the competition for the right to seek oil and, having found it, to produce it under conditions of an agreement with the host country which made the tax-paid cost of the oil almost inevitably higher than that for alternative supplies of oil – an identical procedure, of course, to that which had previously been adopted in France's relationship with Algeria.

In many cases the exploration and development programmes continued to proceed as planned, with the French expenditure of about $300 million in 1965 on exploration for oil to be at least double that amount in 1975. However, not even the considerable political and economic attractions of the special relationships were sufficient to achieve the end which France sought in Iraq, which, preceding even Algeria, was the first Arab country to show that it perhaps did not, or could not, differentiate between the economic exploitation of its resources by France on the one hand and by other capitalist nations on the other! The setback to French policy aims arose in 1968, when Iraq did not accept French proposals for the exploitation of the North Rumaila oilfield, which had been expropriated from the Iraq Petroleum Company a couple of years earlier. Iraq preferred instead to go it alone with help from the communist world, a pattern soon to be repeated, as indicated above, in the case of Algeria.

There is thus good cause to suggest that a special relationship oil policy, even within the framework of a very deliberate general foreign policy orientated to achieving such an aim, is not particularly easy and straight-forward. This is especially so in a situation in which the countries with which the relationship is being sought are not necessarily only concerned with getting the best possible economic return from their oil or, and even more frustratingly, with pursuing a policy which is conditioned by the same logic and consistency as that of the seeker of the relationship. This is seen clearly in the case of Iraq, where, as indicated above, the French proposal to develop North Rumaila was turned down in favour of letting the Iraq National Oil Company (I.N.O.C.) do the work itself. This was in a situation in which there was doubt if the I.N.O.C. had either the technical or the financial capabilities at that time of undertaking the field's development. In this case the offer of assistance by the Soviet Union turned Iraq away from a special relationship with France to an even more special relationship with the U.S.S.R. as part of the new alliance between the Soviet Union and Arab countries following the 1967 Middle East war.

We must recognize the ability and willingness of the oil-producing countries unilaterally to change the terms of agreed special relationships (when they find themselves in a stronger bargaining position either as a result of alternative offers, as in the case of Iraq, or simply because their oil has achieved a higher value, as in the case of Algeria), as a fundamental barrier to the success of such a strategy by oil-consuming countries. This is because such change eliminates the essential rationale of the special relationship policy, viz. ensuring a high degree of security of oil supply. French experience was thus not a particularly happy precedent on the basis of which to persuade the rest of Europe that a special relationship strategy was necessarily going to be of much help in the medium- to longer-term search for the greater security over its oil supplies from O.P.E.C. countries. This was especially true because for Europe as a whole (and for particular parts of it such as the E.E.C.), there was much less political justification than in the case of France for thinking that there were any very good reasons why the oil-producing countries should ever seriously consider the idea of a special relationship, except as a short-term expedient on their part in order to secure a greater share of the profits to be made out of oil production at going prices, or as a means of pushing prices to higher levels so that the opportunity for the producing country to make enhanced profits could be significantly improved.

The producing countries' bargaining position in the 1975–6 search by consuming countries for special relationships with them was, in fact, even stronger than the argument above suggests, for there was little similarity in the 1975 situation in international oil with that which existed at the time when the French 'special relationship' policy was first evolved. When France was seeking its special relationships there was, first, a weak international oil market which gave producers an incentive to consider alternative arrangements for marketing their oil; and, second, there was not much by way of competition from alternative offers, thus putting France in a buyer's market for such deals. But by 1975 the producers had established a controlled supply of crude oil such that there was competition for buying 'special arrangements'. A seller's market had been established in the commodity, so minimizing the chances for buyers to achieve security-of-supply arrangements in this way.

There was also another important consideration relating to special relationship policies. This is the necessary effect of the results of such relationships on national energy policies. As already pointed out above, the achievement of a tied source of crude oil in large quantities, with the intention of making it, at least on the part of the buyer, a more or less permanent arrangement, implies a domestic energy policy which assures refining capacity for the crude oil concerned and markets for its products, in competition both with other oil and with other energy sources. In France this meant the continuation, and the strengthening from time to time, of the long-established system for controlling the importation, refining and distribution of oil. The French companies largely responsible for implementing the special relationship in terms of the flow of oil had to have certain shares of the domestic market reserved for them. Other companies with refining capacity in France then had to accept special relationship oil as part of their throughput, all necessarily leading to a supervised and ordered marketing system in which 'competition has been sluggish at best and the government has kept prices high' (Adelman).

Not, of course, that there is anything inherently unreasonable or necessarily unacceptable about such a system, providing one accepts the basic underlying economic and political organization philosophy that it implies. One should note, moreover, that in the more recent period of rising oil prices on the world oil market, arising from the control over supply now exercised by the producing countries and companies, such a system could have given consumer prices which were no higher, or even lower, than those in non-ordered and non-government-controlled marketing systems. Indeed, the differentials between traditionally higher oil product prices in France and the lower prices in competitive markets like West Germany were gradually eliminated in the period between 1971 and 1976.

At the European level, however, the philosophy of E.E.C. policy (even among the 'Six' and later even more so among the 'Nine', with Britain's weight thrown in in favour of competition) remained orientated towards the idea of 'competition in the market place' as far as energy was concerned. One must, however, note that this philosophy of oil pricing stood in danger by the mid-1970s of becoming even more of a façade than it had been previously under the E.E.C.'s régime of protection for particular fuels. This was because the oil companies responsible for producing and selling most of Europe's energy had already determined to collaborate with each other internationally in order to achieve higher rates of profitability on their activities. Nevertheless, it was at least possible to maintain the façade of competition in the market place, as long as nothing was done to change the fundamentals of Europe's oil supply position. By comparison, the establishment of special relationships with specific oil-producing countries in the French style necessarily implied the extension of the French system of organization of its domestic oil sector to the rest of Europe, and this could have caused as serious disagreements between the member countries of the E.E.C. as they had faced previously in evolving a European agricultural policy.

Thus, given the realities of the world oil situation by 1975 on the one hand, and the European-level market economy philosophy towards the energy sector on the other, the idea and the implications of special relationships between oil-consuming countries or regions and oil-producing countries or regions, which aimed, in the short term, to supplement and, in the longer term, to replace the organizational framework by which Europe's oil had been provided largely by the international oil companies, did not appear to be particularly relevant to the problems of Western Europe's future oil supplies.

Thus, in 1975 and 1976 neither the American-sponsored strategy, based on undermining the O.P.E.C. cartel by means of confrontation through an alliance of consuming countries, nor the French strategy, requiring successful long-term special relationships with the existing important oil-producing and exporting countries, produced very effective results. The French view of the inherent strength and stability of the O.P.E.C. cartel certainly stood up to much closer examination than the alternative hopes of the U.S. and other western nations that they could break up the oil cartel. Though differences in other fields of endeavour and interest amongst the O.P.E.C. countries were clear to see, it was difficult to find any real motivation why any one exporting country would have been prepared to undermine the cartel during the 'heady' period following its successful price increases in 1973–4. Oil-exporting countries, both rich and poor in terms of financial resources, and large and small in terms of their physical potential to produce oil, were hardly able to do better individually than they were doing collectively, for they could all get as much, or as little, money as they wanted by modest adjustments to their volumes of production and/or by large adjustments to their prices. Neither pleas nor threats from the western world (unless backed by the ultimate threat of the use of military forces) could easily alter this situation for individual explorers at a time when the solidarity of O.P.E.C. had produced so many benefits for its members.

The best hope lay in persuading the ruler of a very Islamic Saudi Arabia that it was his duty to save the western economic system which, in large part, had emerged out of the Judaeo-Christian tradition. Stranger things have perhaps happened in history, but not very often. Or was the ruler of the world's largest oil reserves expected to save the western world's economic system in his own interests? This presupposes that the interests of Saudi Arabia and other oil producers were generally identifiable with those of the west. There was, however, no real justification for such a view at a time when the world was emerging from a period of over 400 years in which the west had kept the benefits arising from its economic system very much to itself! And if the west's 'threats' to make itself much more self-sufficient in energy as soon as possible were successfully carried through, then would the result really be an adverse one for the O.P.E.C. countries? By the 1980s, given continuing large oil sales at high prices in the meantime, they stood to accumulate enough funds to be able to live off the interest, and thus then have a minimal interest in selling oil anyway. Such a view was, indeed, clearly stated time after time by another leader of an O.P.E.C. country, the Shah of Iran, who expected his country to have a set of non-oil interests by the 1980s which collectively would be more important than Iran's oil interests. Nevertheless, had the American strategy of breaking O.P.E.C. succeeded it would, at least, have assured the western world of its required flow of energy by the 1980s. In the meantime the problem would have been to ensure that the system survived.

On the other hand, the French were much more correct in their evaluation of the inherent strength of the O.P.E.C. system. However, each additional step taken by France (and others) to seek a special relationship with the member countries of O.P.E.C. strengthened the hand of the cartel, without effectively doing anything to ensure a flow of oil (given the continuing ability and the possible motivation of the producers to reduce production at will and, moreover, their further ability to turn off their need for further technological help at any moment in time). At the same time the French attitude served to divert attention and resources away from the need to take action at home to solve the problem. The action required in this respect involved, first, the need by the west greatly to reduce its use of oil and, second, the immediate and pressing need to stimulate and accelerate the production of indigenous oil and other alternative energy sources. Both psychologically and in physical terms a strategy based on the idea of special relationships between O.P.E.C. nations and oil-consuming countries inevitably undermined the efforts necessary to achieve such aims.

Thus, by 1978, neither special relationships with O.P.E.C.'s members in the way favoured by France, nor the break-up of the organization sought and expected by the United States and most of its western allies, had been achieved. Indeed, the high degree of solidarity to the member countries of O.P.E.C. steadily became increasingly obvious to, and reluctantly accepted by, the western industrialized nations. Thus, their earlier hopes, described above, that strains and stresses amongst the world's oil producers over questions of price, and over levels of production would lead to a break-up of the organization in the near future had to be discarded. Instead, the fact of the oil-producing countries domination of the international oil system was largely accepted, with all that that implied for the future organization and development of the non-communist world. As a consequence there finally emerged a much greater awareness on the part of most oil-importing nations, rich and poor, of the urgent need to reduce their dependence on O.P.E.C. oil.[2] We shall return later in the chapter to these considerations, in the context of a discussion of a comprehensive alternative strategy which involves full and formal cooperation between O.E.C.D. and O.P.E.C. as a mechanism for solving not only the continuing problems of the supply and price of oil, but also the international economic problems which continue to beset the western system, as a result in part of the changes in the world of oil power since 1970 and of the uncertainties to which these changes have given rise. Before proceeding to this strategic question, however, there was one important tactical development in the world oil situation in the later 1970s to which we must devote some attention. Paradoxically, in light of the anti-special-relationships policies of the U.S., in respect of the efforts of individual industrialized countries such as France and Japan to close bilateral deals with individual oil-exporting countries in the period up to 1976, this concerns a type of special relationship which was established between the United States and Saudi Arabia.

Saudi Arabia had become, as shown in Chapter 4, far and away the single largest oil-exporting nation. Its exports by the late 1970s accounted for over a third of all the oil which moved in international trade. It was also a country which had a wide range of options in deciding how much oil to produce. On the one hand it was able, if it so wished, to reduce its production to no more than 25 per cent of its 'normal' level of output of about 500 million tons per year. It simply did not need the money which its oil exports could earn. It could thus initiate action severely to limit its exports, the effect of which would be greatly to strengthen O.P.E.C., and oil prices, as oil demand came to exceed the available supply. This was an option which the United States had to try to persuade Saudi

2. The author elaborated such a strategy for Western Europe in the Stamp Memorial Lecture he was invited to give before the University of London in 1975, *The Western European Energy Economy: Challenge and Opportunities* (Athlone Press, 1975). Unhappily, the diagnosis and the prescriptions were not accepted or, indeed, acceptable. The continued failure to respond to O.P.E.C.'s challenge extended the period of the west's self-inflicted wounding arising from the oil crisis.

Arabia not to follow, by using every possible opportunity to point out the dangers to which this would give rise. It did this particularly by drawing attention to the 'radicalization' of politics to which a greatly strengthened O.P.E.C. could give rise, in both general international terms and in respect of policies in some of the O.P.E.C. countries themselves. Given this, then the chances of survival of the capitalist western system would be much reduced, to, it was argued, the disadvantage and even the danger of the Saudi Arabia régime. In brief, Saudi Arabia's best interests would not be served by action to reduce its supply of oil, as this would undermine the viability of régimes friendly to the existing Saudi Arabian system of government.

On the other hand, Saudi Arabia had the physical ability to increase its output of oil by 50 per cent or more. This was a prospect which would so enhance the total world supply (outside North America and the communist countries) that the possibility of O.P.E.C. surviving the consequential imbalance between supply and demand would be very much in question, unless all or most other O.P.E.C. members were ready, willing and able to reduce their output to offset the Saudi Arabian increase in production levels. Such counteractions to the Saudi challenge would have been difficult for most of O.P.E.C.'s members to sustain for longer than a few months, because of their need for a continuing flow of revenues and of foreign exchange earnings from oil exports to keep their ambitious development programmes moving ahead. Thus, the United States tried to persuade Saudi Arabia to commit itself to increasing its oil output. An additional production of a million barrels of oil per day (50 million tons per year) in the short term was seen as likely to make a significant enough difference to the overall international supply/demand situation that the ability of O.P.E.C. to maintain, let alone to increase, its prices would be undermined.

Over the longer term the United States, through the intelligence that was available to it via Aramco, the American company which had, in effect, been the Saudi Arabian oil industry, calculated that there was a potential for a 60 to 100 per cent increase in Saudi Arabian production. Such an expansion of Saudi production could, by the mid-1980s, provide the United States with all or most of the additional oil imports which the United States then thought it would need by that time and so give the country a high degree of security in respect of its oil supply. It would, of course, also have undermined the expectations of O.P.E.C. in the late 1970s that rising import demands by the United States, coupled with rising demands for O.P.E.C. oil from other parts of the industrialized world, would secure the future of the organization by keeping demand running comfortably ahead of the quantities of oil which the O.P.E.C. nations needed to supply to maintain the flow of oil revenues required for financing their economic development programmes.

Other O.E.C.D. member countries were not involved in this United States initiative with Saudi Arabia. Indeed some of them would have found such involvement embarrassing because of the anti-O.P.E.C. undertones which were implicit in the policy at a time when they were trying to develop closer economic and political ties with other O.P.E.C. countries: for example, France with some of the Arab producing countries, and the United Kingdom with Venezuela. Nevertheless, the tactic had much tacit support throughout the group of industrialized countries in that it was seen as the best way to avoid a further intensification of the oil supply and price problem in the medium term. As such, it was thought to offer a chance of eliminating one of the most important uncertainties which the O.E.C.D. world thought that it collectively faced in respect of its development prospects in the 1980s.

The possible containing of O.P.E.C.'s influence in this way was, moreover, seen to have two further advantages. One of those was that it was an inherently less dangerous option than alternative ways which had been considered in the past. These had included possibilities such as military action against appropriate targets in the Middle East, and the initiation of general economic sanctions against the member countries of the organization. Both would have incalculable repercussions in the Third World, and in terms of east–west relationships – and so they represented high-cost options. Such results would not have arisen from the Saudi Arabian–United States arrangement. It would have had the appearance of being a strictly bilateral deal which other countries could simply ignore. It could also have been presented in strictly commercial terms, as part of new arrangements for the ownership, production and marketing of Saudi Arabian oil in the context of a changed role in Saudi Arabia for Aramco. In such circumstances Third World countries would have had no need to approve it or even to have commented adversely upon it, especially as its effect would have been to reduce the price they had to pay for their oil imports.

In tactical terms, therefore, oil power play in the later 1970s revolved around the interrelationships of the United States and Saudi Arabia. The play, moreover, took place in the context of an apparent belief, not only of the two parties themselves, but also of the governments of many other countries including those of most of the countries of the industrialized world, that an effective *rapprochement* between them was necessary in order to ensure that the otherwise inevitable disequilibrium in the non-communist world's oil supply and demand situation could be postponed for at least some years and, more hopefully, be avoided altogether as alternative energies were brought in to substitute the use of oil. In other words joint U.S./Saudi Arabian regulation of the oil market came to be seen as the essential element in a strategy for eliminating possible excessive demands by other O.P.E.C. members for yet higher oil prices in the first half of the 1980s.

Satisfactory progress in the negotiations appears to have been made up to mid-1978 when the prospects for relative oil price stability at levels of about $15 per barrel seemed good. The beginning of the revolution in Iran, however, culminating in the effective overthrow of the Shah's régime in December 1978, undermined these prospects. The consequential elimination of Iran as a major oil producer and exporter led to a perception of severe imbalance in international oil supply and demand conditions. There was thus an immediate re-emergence of speculative trading on the world oil market. The result was 'the second oil price shock', viz. the further doubling of oil prices to about $30 per barrel by early 1980 when the market, under the influence of increased Saudi Arabian production and a decline in demand, started to settle down once again. This result of U.S./Saudi Arabian cooperation augured well for the future (in as far as this could be true with oil at $30 per barrel), given that stability had been achieved in the market even in the absence of Iran as a major exporter, and again the prospects for equilibrium in the world oil market and stability in oil prices seemed to be much better. Indeed, by the summer of 1980 there was an excess supply of up to 3 million barrels of oil per day (= 150 million tons per year) on the international market and this helped the prospects generated by Saudi Arabian/U.S. action and agreements. Some agreements, such as the one on the future of Aramco, were public, whilst others, which were part of a wider agreement between the two countries, were secret (such as an apparent Saudi guarantee to ensure certain minimum levels of oil availability to the United States).

Another event in the world of other oil-exporting countries occurred, however, a few months later and threatened to upset the situation once again. This was the Iraqi attack on Iran in September 1980. This proved to be disastrously unsuccessful and the Iranians were able to counter-attack very quickly. The Iranian's advance closed off Iraq's access to the Gulf and lead to the almost immediate halting of oil exports from Iraq, which, in the aftermath of the Iranian revolution's closure of most of the Iranian oil industry, had become the world's second largest oil-exporting country (after Saudi Arabia). The elimination of most of Iraq's exports (which had been more than 3 million barrels per day) instantly eliminated the excess supply over the level of demand in the autumn of 1980 and thus opened up the possibility of yet another scramble for oil with the prospect that oil prices would be levered up to still higher levels.

Thus, the reality of oil power at the beginning of the 1980s still related to the issues which had been raised since 1970 by O.P.E.C.'s challenge to the established international economic order: that is, to issues arising from the transfer of resources from the industrial to the oil-producing and exporting countries, and to the relative improvement of the latter's position in the hierarchy of the world's nations. There was also continued perception of uncertainty over the supplies of oil which O.P.E.C. countries were willing to make available to important countries: in the context of which the richest and most powerful nations could increasingly seek to ensure their requirements were met, at the expense of the poor oil-importing countries of the Third World. Thus, in 1981 the most widely-accepted scenario for oil in the 1980s still saw the readiness and the ability, or even the willingness, of O.P.E.C. members to continue to deliver enough oil to the world market to avoid supply problems as the major problem in the international oil situation: and with a very widely held expectation that the price of oil would continue to increase to $60, or even eventually to $100 per barrel or more.

Such fears persisted in spite of the fact that the continuing Iraq–Iran war, which severely reduced both countries' oil output and exports, did not lead to a supply problem. On the contrary, by late 1982, in spite of the restraints on oil production from two of O.P.E.C.'s previously most important exporters, coupled with a tentative agreement among all O.P.E.C. members (except Saudi Arabia) to limit their levels of production, and Saudi Arabia's severe cutback in its production to about 6 million barrels per day (compared with over 10 million in 1980), the essence of the world market was already one of weakness. Too much oil was chasing too few markets and there were no immediate prospects for change in the situation in the absence of any further dramatic political development in the Middle East. More important still, the fears persisted even though they were unrelated to the emerging fundamentals of longer-term international oil supply and demand relationships. Instead there was a mistaken belief in the western world that oil is an inherently scarce commodity because the world resources are very limited and are being depleted very quickly. This belief then generated behaviour which not only helps to create the propensity for the development of a longer-term scarcity (by reducing the motivation to search for oil), but which also persuaded member countries of O.P.E.C. to behave as though oil is inherently scarce, so that their importance in the world system was ensured. O.P.E.C. continued for some time to see the world of oil in this way in spite of the weakening oil market after 1981. The oil-exporting countries assumed the weakness to be the result of a short-term imbalance between too little oil demand at a time of economic recession to match the supply potential. It was not, however, only O.P.E.C. and its members which argued along those lines. The I.E.A., the 'club' of the western world's industrial nations which was originally launched in 1974 by U.S. Secretary of State Kissinger to combat and undermine O.P.E.C. continued to argue the same line even more strongly as, for example, in its 1981 analysis of the long-term prospects for world energy (*World Energy Outlook*, O.E.C.D., Paris, 1982) and even in more recent documentation (in its 1985 *Energy Review*) in which the emphasis is still placed on the essentially temporary nature of the weak oil market. In this way O.P.E.C.'s continuing and future status in the world of oil power has been exaggerated beyond that which is justified by the now rapidly emerging longer-term global energy and oil prospects.

Since 1979 the demand for oil has ceased to grow in the world outside the communist countries while the contribution of O.P.E.C. oil to the total supply has fallen very sharply. This is shown in Table 3. In 1985 the O.P.E.C. countries exported less than half of the amount they exported in 1973 while the 1985 use of oil in the non-communist world (outside the O.P.E.C. countries themselves) was about 15 per cent lower than it had been in 1973 (after rising to a peak in the meantime – in 1979 – when it was 5 per cent up on the 1973 level). These global figures do, of course, mask important regional variations in the production and use of oil. The O.P.E.C. countries themselves increased their own consumption of oil very markedly; there was a somewhat increased use in the non-O.P.E.C. developing countries: in the United States there was a decline in use in spite of the fact that oil had to substitute for a declining availability of natural gas; and there was a sharp fall in the use of oil in both Western Europe and Japan. Overall, however, there has since 1973 been a marked change in the shape of the oil consumption curve. Prior to 1973 this had showed an exponential growth of about 7·5 per cent per annum for almost twenty-five years. It then flattened to less than 1·5 per cent per annum between 1973 and 1979 and since then growth in use has been replaced by a decline rate – at an average over the six years, 1979–85, of about 5 per cent per annum.

In brief, the combination of the impact of much higher oil prices and of other factors, such as government efforts to control the use in many countries, has been severely under-estimated in analysis of future oil

Table 3. Sources of energy used in the non-communist world (excluding the O.P.E.C. countries in 1973 and 1985

	1973 mtoe*% of total		1985 [est.) mtoe*% of total	
Total Energy Use	4,045	100	4,300	100
of which				
a) Imports of O.P.E.C. Oil	1,480	36·5	700	16·3
b) Other Energy Imports†	100	2·5	250	5·8
c) Indigenous Production	2,465	60·9	3,350	77·9
of which				
i) Oil	760	18·8	1,225	28·5
ii) Natural Gas	765	18·9	775	18·0
iii) Coal	805	19·9	1,025	23·9
iv) Other	135	3·3	325	7·5

* mtoe = million tons oil equivalent

† Oil, natural gas and coal from the Centrally Planned Economies

needs. The effect of this is made even clearer if one compares oil use in 1985 with what was expected to be used in 1985 in forecasts made immediately prior to the first oil 'crisis' of late 1973. According to those forecasts, world production (and use) of oil in 1985 should have been more than twice that of 1973. Instead, even after taking the continued increase in the use of oil in the communist world into account, it is less than in 1973. Put at its simplest one can say that only one barrel of oil is now being sold for every two that were expected to be sold by this time.

Conventionally, the failure of oil demand to expand is related most strongly to 'poor industrial and economic performance' and there is, very obviously, something in this argument, given the relative weakness in most western economies since 1973. But in that the gross domestic product of all industrial countries has grown in the intervening period, albeit at a lower rate than prior to 1973, there is much more to the change in the shape of the oil demand curve than simply poor industrial and economic performance. This becomes clear when we look at Figures 3, 4 and 5. Figure 3 shows the relative importance of reduced economic growth, on the one hand, and of increasing efficiency in energy use on the other in keeping the industrialized world's total use of oil at about the same level as in 1973: the diagram indicates how the impact of the second factor has become gradually more important over the years (as energy conservation efforts have become increasingly effective) so that by 1984 'savings' accounted for almost half of the energy that had not

251

been used (compared with the use there would have been against 1973 norms). Figure 4 then indicates how almost every country in the industrialized group of countries has improved the efficiency of its energy use since 1973. The overall average improvement by 1983 was about 22 per cent but the fact that a number of countries – notably Japan – have done much better than this indicates that the process elsewhere still has a long way to go. Figure 5 separates out the evolution of industrial world oil use from energy use – and relates both to the process of economic growth. From 1965 to 1973 energy use and G.D.P. (Gross Domestic Product – a measure of output in an economy) grew in a roughly 1:1 relationship with each other: over this period, however, the increase in the use of oil was much faster indicating the growing oil intensity of the development process in the industrialized countries. Since 1973 one can again see the divergence of growth in energy use from economic growth but of even greater importance is the way in which the use of oil has declined even more – to below its 1973 level. Thus, by 1984 when G.D.P. in the industrialized countries stood at over 125 per cent of its 1973 level, the use of oil was only 80 per cent of the amount used in 1973.

In the light of this evolving pattern of experiences in respect of recent changes in the use of oil it looks as though the kinds of forecasts that have been and, indeed, are still being made about the future needs for oil in the world economy are probably much too high. Reports published in the late 1970s, such as the *World Alternative Energy Strategies* report, the World Energy Conference report on *World Energy Resources* and the Central Intelligence Agency's report to President Carter, all worked with oil-use growth rates of 5 per cent or more and so produced oil demands for the future which indicated pressures on the ultimate potential world supply of oil by the mid– to late 1980s. From such analysis they concluded that there must be continued rapid increases in the price of the commodity. Such increases were presented as being inevitable arising from the physical shortage of oil in the world, rather than the result of O.P.E.C.'s actions in restricting supply to less than the anticipated demand. More recent studies, notably the I.E.A.'s *World Energy Outlook* (1982) and, even more so, its *1984 Energy Review* (1985), have had to take the falling use of oil into account. Nevertheless, they have continued to forecast that oil use could not fall much further (if any at all) and that growth in its use will be re-established as the norm by the late 1980s. By the year 2000, the I.E.A. forecasts, world oil use will be up to 50 per cent higher than in 1980 (and thus about 65 per cent higher than the lower use in 1985). There are several reasons why such growth rates in the use of oil now appear to be very optimistic.

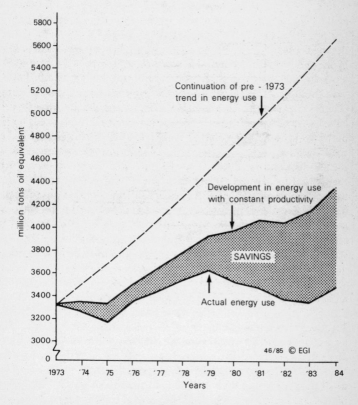

Figure 3. I.E.A. countries' energy use 1973–84.

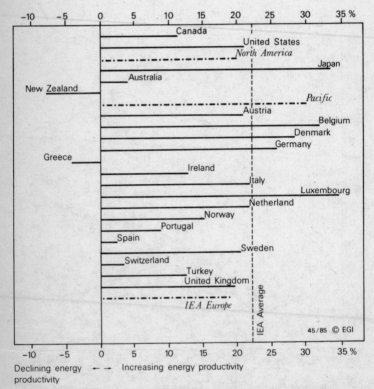

Figure 4. Changes in energy productivity in individual I.E.A. member countries 1973–84.

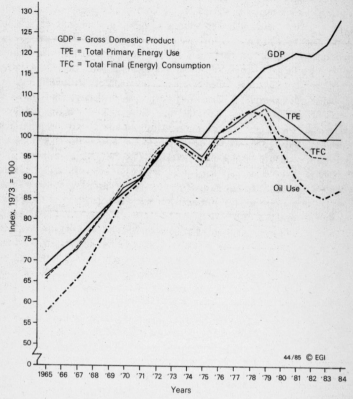

Figure 5. The decoupling of economic growth from energy and oil use 1965–84 in the I.E.A. countries.

First, they are based on optimistic assessments of the western world's economic development potential. The problems of readjustment following the politico-economic difficulties created by the oil revolution in 1973 remain very much in evidence. These include problems such as high rates of inflation, balance of payments difficulties, monetary instability, unemployed resources (especially labour), and the failure of international trade to grow as expected. Though efforts are being made to define these problems and to find solutions to them, these have not yet been generally successful and in some respects things still seem to be getting worse. Nevertheless, in many countries of the industrialized world it seems to be politically impossible to take such difficulties into account in forecasting future economic growth rates, for fear of the consequences if unemployment were shown as likely to continue to rise from its present high levels. As a result, over-optimistic views on economic expansion appear to be written into forecasts of the future needs for energy in general, and for oil in particular. The forecast use of oil is especially seriously affected by such considerations, as many forecasts tend to have fixed, or nearly fixed, supplies of other energy sources written into them (because, for example, it has simply been decided that so many million tons of coal will be used). Thus the expected use of oil is highly variable relative to the degree of development incorporated into the planning outlook. If the forecasts for development are optimistic, then the expectation of future oil use must necessarily be even more optimistic.

Second, many forecasts for future energy use are still using a high energy/G.D.P. coefficient (this measures the relationship between the rate of increase in energy use and the rate of increase in the level of economic activities). For example, the Commission of the E.E.C. was, until 1983, still working with a coefficient of 0·8: a ratio, that is, which is almost as high as the one which was found to apply in Western Europe during the period of falling real energy prices up to 1973. The I.E.A., in its 1982 *World Energy Outlook*, noted the recent fall in the coefficient (to an average for the O.E.C.D. countries of 0·34) but then proceeded to its forecasts on the assumption that the coefficient would move back to a level of at least 0·63 by 1985 and up to as high as 0·81 by the year 2000. Given the much higher energy prices, and the way in which these have persuaded consumers to be very much more careful in the way they use energy, the use of an hypothesized return to a high coefficient does not seem to be justified for estimating future energy needs. Even without further energy price rises the value of the coefficient may be expected to fall still further as the implementation of energy-saving technology and changing societal organization to more energy conservationist practices continues to diffuse – to increasing numbers of energy users in the industrial countries (where the learning process in respect of conservation is already well under way) and to users in the developing world. The process in the latter will inevitably be slower than in the industrialized countries because of the lack of investment funds, of capital and of trained personnel but the results, when achieved, will be even more dramatic than in the industrial countries for two reasons: first, because the energy coefficients are much higher (an average of over 1·0 as shown in Figure 2 on p. 168) and so have further to fall; and second, because the developing countries are relatively more dependent on oil so there will be a greater relative effect on this energy source. In any case future energy demand will be less than is currently expected in official forecasts and, for the reasons given in the previous paragraph, the negative influence on oil use will continue for the forseeable future.

Third, too little attention is being given to the possibilities of increased use of alternative energy sources in many countries of the world, especially amongst the industrialized nations. In Canada, for example, there is a potential availability of large volumes of new natural gas supplies. Gas thus seems likely to replace oil in many uses, especially as pipelines are built to areas where natural gas has not previously been available (in, for example, the eastern part of Quebec Province and in New Brunswick). The same is true for Australia, where a transcontinental pipeline system is now being built to take gas from the north-west Australian continental shelf to the energy-consuming areas of the south-west and the south-east of the country. Of even more importance is the likely increase in the availability and use of natural gas in many countries of Western Europe. The expansion of the supply of natural gas in Western Europe over the last decade or so was described in Chapter 5, and major new resources have recently been found in the northern part of the North Sea basin and further north off the coast of middle and northern Norway as well as to the west of the Shetlands. As shown in Map 6 (p. 130), these resources will justify the building of major new pipelines to make more gas available in many of the continent's main energy-consuming areas and where it can be expected to continue to substitute the use of oil products. At the other end of the continent a new pipeline for natural gas has been built from Algeria to Italy across the Mediterranean and it has already been decided to increase the capacity of the line. This will be followed by a second trans-Mediterranean line to south-eastern Spain. Such developments will help to diminish existing markets for oil in much of southern Europe, where, hitherto, there has been little competition for oil from other energy sources.

Meanwhile, even in the United States there has been a marked reversal in the previous downward trend in the availability of natural gas. The potential supplies for this change in the rate of discovery of gas reserves were known to be available and the industry simply waited for the government to modify its control over gas prices (for gas pipelined over state boundaries), so that prices could be raised to levels which made increased production sufficiently profitable. At the same time the U.S. coal industry has been and continues to be expanded and encouragement is being given to the electricity industry and other industries to use coal instead of oil for steam-raising purposes. Natural gas and coal between them are thus squeezing the markets for oil in the United States and oil consumption has fallen sharply since 1979. This fall is particularly signifi- cant in influencing the overall non-communist world pattern of oil use, given the high relative importance of the use of oil by the United States (it currently uses almost 40 per cent of oil used in the world outside the communist countries) and also the fact that it was only the growth of the demand for oil by the U.S.A. between 1973 and 1978 which caused the overall non-communist world production of oil to increase by a few percentage points. If oil demand in the U.S.A. continues to be restricted, as since 1979, then there will continue to be a significant effect on the world oil supply/demand situation.

Fourth, too much emphasis is being placed on the growth of oil demand in the less developed countries. We have shown previously (in Chapter 7) just how dependent most of these countries are on oil and how quickly their oil demands increased as they went through the initial stages of industrialization and development. This process, prior to the upset in the world economy in 1973, was expected to continue but since then the economies of most developing countries have been under severe strains and the outlook for them remains far from bright as a conse- quence of the depressed international economic situation. Thus, their use of oil will now, as indicated above, grow much less quickly than expected. In addition, these countries also have another powerful motivation to conserve energy – to reduce their imports of oil which have come to constitute a massive burden on their balance of payments. In the short term they are, in many cases, simply having to do without the oil they need to import to maintain their industrial and transport sectors, but they are now also developing alternative, domestic energy resources and thus also contributing to the down-turn in the growth rate of oil demand.

Thus, official expectations of future global growth rates for oil as high as 3 to 4 per cent per annum now seem to be misplaced. Given an outlook for economic growth which will be both spatially and temporally variable (some countries will do better or much better than others, and some years will be better or much better than others), and given the encouragement of the production and use of alternative sources of energy, then it appears most reasonable to expect an average per annum change in the rate of use of oil which will be only just positive at best; it may even be negative, or strongly negative, over the rest of the decade and even through the 1990s. This would not, of course, be a reasonable conclusion if the price of oil were to collapse, so that consumers no longer had to worry about its cost and so that alternative energies could no longer compete. From the perspective of 1985 the possibility of such an oil price collapse cannot be excluded but this is an issue to which we shall return later in the chapter. Neither would it be reasonable if there were to be a massive revitalization of the western world's economy. This, however, could only arise from a reversal of politico–economic policies which are being followed in the United States and other leading industrial countries, and from a significant easing of the present difficult economic environment for the developing countries. Neither of these seems likely and a combination of both is very unlikely. Thus massive economic revitalization does not appear to be in prospect.

Viewed in this context of a low growth rate in the development of oil demand there are no necessary medium-term oil supply problems other than those of a purely political character. This is revealed in the contrast between existing producing capacity, most of which is in the O.P.E.C. countries, and the levels of demand that are currently being made on it. O.P.E.C.'s total theoretical production capacity rose in the 1970s to a peak of 37 million barrels per day (= 1,850 million tons per year). In 1979 O.P.E.C. members produced more than 30 million barrels per day (about 1,500 million tons in the year) to achieve a production level which was close to that technically achievable over a year. Less than a year later, however, O.P.E.C. had to try to find ways and means whereby they could cut back their levels of production to 23 million barrels per day (= 1,150 million tons per year) and even then failed to find markets for all the oil they expected to produce. The situation continued to deteriorate throughout 1981 (with O.P.E.C.'s average daily production down to 22·5 million barrels and at least 1 million barrels per day below that level by the end of the year). Discussions then began in earnest on the establishment of country quotas for O.P.E.C. members, with 17·5 to 18 million barrels per day as the overall objective for total production. By the middle of 1982 production was down to that level, but only because Saudi Arabia and a number of other Gulf producers cut back below their quotas in order to overcome the problem of production above their quota rates by Iran and Libya. Since then the situation has deteriorated even further so that in 1985 not even a total production by O.P.E.C. members as low as 15 million barrels per day (equal to 750 million tons per year) could avoid an element of over-supply on the market. Moreover, O.P.E.C.'s output was restricted to this level only by the willingness of Saudi Arabia to allow its own production to fall well below its quota of 4·3 million barrels per day (perhaps even to little more than half this level). Clearly O.P.E.C.'s strategy and cohesion has been called increasingly into question by the low demand levels for oil compared with the volumes which the oil-exporting countries thought they needed to export, for economic and/or strategic reasons. Even in some non-O.P.E.C. countries there has been the emergence of some under-utilized oil-producing capacity: as, for example, in the oil-producing Canadian province of Alberta, and in Alaska as a result of transport difficulties. In addition, O.P.E.C. has tried to persuade other major non-O.P.E.C. producers such as Britain, Mexico and the Soviet Union to cut back their levels of production and/or exports – though with only modest success as these other oil-exporting countries have all had powerful motivations to expand their hitherto limited roles in the international oil market. Table 3 (p. 251) summarizes the quite dramatic changes which occurred between 1973 and 1985 in the contributions of different sorts of energy to total demand. In particular note how the contribution of O.P.E.C. oil to the total fell from 36·5 per cent to only 16·3 per cent while oil produced in non-O.P.E.C. (and non-communist) countries increased its contribution from 18·8 to 28·5 per cent. O.P.E.C.'s domination of the market has been much diminished. It can also be calculated, however, that the 55 per cent share of O.P.E.C. plus non-O.P.E.C. oil to total energy supply in 1973 had fallen to 45 per cent by 1985 reflecting the concern over the previously very important role of oil in the non-communist world's energy economy – and the need to reduce it in the light of events since 1973.

This new weakness in the overall oil supply/demand situation also affects new oil-producing capacity which is being, or which could be, developed based on the marked improvement in the proven oil reserves situation over the last fifteen years. The development of oil reserves is shown in Table 4. In eleven of the fifteen years more oil was added to reserves than was used and in two of the other years (1977 and 1979) the net excess of use over additions to reserves was only just negative. Over the whole period $334·3 \times 10^9$ barrels were used while almost $509·7 \times 10^9$ barrels of oil were added to reserves – put more simply, for every

Table 4. Annual additions to reserves, oil use and net growth/decline in reserves, 1970–85

	Reserves at Beginning of Year	Use of Oil in Year	Gross Additions to Reserves in Year	Net Growth (+) or Decline (−) in Reserves
	(in barrels × 10⁹)			
1970	533	17·4	62·4	+45
1971	578	18·3	40·3	+22
1972	600	19·3	−3·7	−23
1973	577	21·2	35·2	+14
1974	591	21·2	32·2	+11
1975	602	20·2	31·2	+11
1976	613	21·9	3·9	−18
1977	595	22·6	15·7	−7
1978	588	22·9	44·9	+22
1979	610	23·7	21·7	−2
1980	608	22·8	33·8	+11
1981	619	21·3	67·3	+47
1982	665	20·1	30·1	+10
1983	675	20·0	21·1	+1
1984	676	20·6	43·6	+23
1985	699	20·8	30·0	+10
Totals	—	334·3	509·7	+177

Sources: *Oil and Gas Journal*, 1970–85; *World Oil*, 1970–83; DeGolyer and MacNaughton's *Annual Survey of the Oil Industry*, 1975–83.

barrel of oil which was used between 1970 and 1985 more than 1·5 barrels were added to reserves so that the ratio between reserves and production is now at an all-time high (except for two or three years in the mid-1950s immediately after the massive up-grading of reserves in the Middle East to a level which represented the reality of the situation rather more closely).

New potential for increased production exists first of all in the O.P.E.C. countries themselves. In Iraq, for example, major new areas of oil reserves have been proven and offer the promise of a potential production level which would be much more than enough to compensate for possible declines in production from the fields which are already developed. Similarly in Venezuela, where oil potential exists in deeper formations in areas already drilled, as well as in new regions – especially the country's extensive offshore areas – and, in the longer term, in the recovery of heavy oil from the vast Orinoco oil belt. Given the demand

261

and given time Venezuela could, if it wished, push its production well above the 2·2 million barrels per day it had hoped to achieve by the mid-1980s (in order to fund its ambitious development plans), and the 1·7 million barrels per day it was in fact achieving at the end of 1985.

Much new potential for increased oil production also exists in countries which are not members of O.P.E.C. Notable in this respect are the prospects for increased production potential from the North Sea, to which attention has already been given in Chapter 5. More than thirty giant fields and over 200 other fields have already been discovered, but most of the latter are not currently being developed. From a technical standpoint they could all be developed. Nor is there any doubt but that the relatively limited amount of investment capital required for their exploitation could be raised. The finance needed is small in relation to what Western Europe has been used to investing each year in its energy sector and even more limited in relation to the total ability of Western Europe to generate investment funds. Even the prospect of the North Sea being capable of producing well over half of Western Europe's declining oil needs year by year for the rest of the century does not exhaust the potential availability of European oil, given that there are very much larger regions of the continent's offshore shelf which still remain to be explored. The number and extent of the potential areas for oil production are shown on Map 10 (p. 263). At least some of these regions will be opened up by new exploration before 2000 and there is a good prospect that one or more of them will contain large quantities of oil. Europe should thus be able to sustain, over a period extending well into the twenty-first century, its recent achievement of a much higher degree of self-sufficiency in its oil requirements than that to which it was used in the period of its increasing degree of dependence on oil in the 1950s, 1960s and the early and mid-1970s. It is within the context of this highly favourable medium-term outlook for oil production in Western Europe that appropriate political and economic decisions are required for achieving not only a continental-wide integrated oil system, but also a pricing system for energy in Europe related to the cost of producing indigenous energy, rather than to the O.P.E.C.-determined price of internationally traded oil. This would ensure that Western Europe is not too seriously disadvantaged compared with the United States, where the domestic economy has generally been protected over a number of years from the international escalation of oil prices, and where there is little doubt that the country will maintain the much higher degree of self-sufficiency in energy: an achievement which it has viewed as one of the cornerstones of its economic policy since the revolution in the world of

Map 10. Western Europe – potential offshore areas for oil and gas production.

oil power in the early 1970s. It has been able to do this, moreover, with energy of lower cost than imported O.P.E.C. oil.

But this very important Western European oil potential is not the only non-O.P.E.C. oil which could be much more intensively and extensively exploited and so help to increase the world's medium-term potential supply. Indeed, every major oil-importing country has been taking a new look at the possibility of indigenous oil production, particularly those countries with extensive areas of continental shelf from which, given the recently much improved technology of offshore exploration and production, large quantities of hydrocarbons are expected eventually to be produced. As Map 11 clearly shows, the non-communist world's potentially petroliferous offshore areas are several times larger than the onshore areas from which, to date, almost all of the world's oil production has come.

In other words the likely absence of an oil supply problem for at least the next twenty-five years is also a function of the extensive potential world oil resource base which remains to be investigated and developed. As already suggested at the end of Chapter 7, this is particularly important in respect of opportunities in the Third World – the world of the less developed countries – as can be seen from Map 11 and even more clearly from Figure 6 (p. 266). This figure shows the relative size of the potential oil-bearing areas in different parts of the world and the amount of actual drilling for oil which has taken place to date in the various regions. So far, it is not much of an exaggeration to suggest that the Third World has been left severely alone as far as oil exploration is concerned and so still represents a potential for oil developments in the medium-term future.

In the light of these three considerations – existing producing capacity, the potential for developing new capacity, and the oil resource base – the oil problem for the future does not seem likely to be one of a physical shortage which, if it existed, would certainly lead to a scramble for oil. Rather it is one in which difficulties in supplying the market would continue to arise if O.P.E.C. were once again allowed to dominate the international oil supply system.

In spite of the surplus production potential for oil over the level of oil demand, the influence on oil supplies by the O.P.E.C. countries, and hence their influence on the price of the commodity, has been largely maintained so that prices, in spite of a series of declines since 1981, have through to the end of 1985 been held way above the level needed to ensure good profits for all producers. There are two reasons for this. The first is an underlying degree of solidarity in O.P.E.C. The organization has managed to hold together in spite of the dramatic change in the

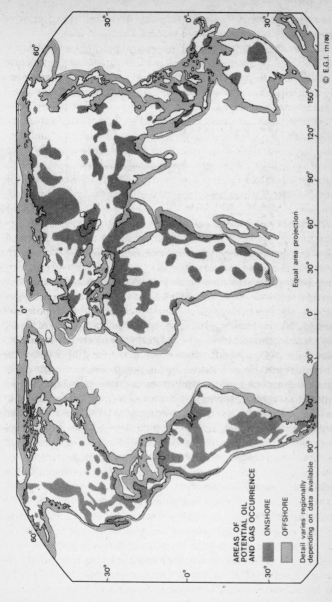

AREAS OF
POTENTIAL OIL
AND GAS OCCURRENCE

ONSHORE

OFFSHORE

Detail varies regionally
depending on data available

Equal area projection

© E.G.I. 171/80

Map 11. Areas of actual or potential oil and gas production: onshore and on the continental shelves and slopes.

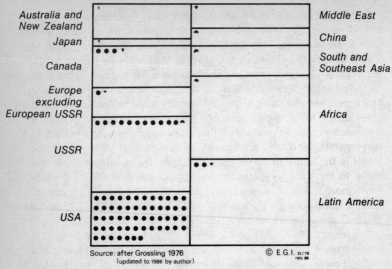

Figure 6. Drilling for oil in relation to size of prospective oil-bearing areas. (Each complete dot represents 50,000 wells drilled.)

demand for oil over the last three years, and in spite of external attempts to exacerbate the differences in view and in approach between its member countries. Following the difficulties between 1977 and 1979 in the organization, as a result of the dual pricing system which emerged when Saudi Arabia and two other members of O.P.E.C. refused to put up their prices as much as the rest of the membership, and the threat to prices from the too high levels of production since 1980, there have been serious attempts to reach agreement on the most fundamental issue affecting the future stability of O.P.E.C. This is the issue of the levels of production which should be 'allowed' for individual member countries, and the ways and means whereby such a prorationing agreement can be monitored and controlled. Success in this respect has so far only been partially achieved (because some member countries have not accepted their quota limitations). Greater success will demand more discipline than hitherto by most member countries. Nevertheless most members of O.P.E.C. have positively sought to reduce their levels of production because they have recognized that they could not hope to maintain the real price of oil at anything like the peak level it reached in 1980–81 without a serious and continuing loss of markets. Thus, production limitations with a gently falling real price may well have become the objectives which O.P.E.C. is now trying to achieve.

O.P.E.C. is being helped – albeit inadvertently – in these aims by the second reason why oil supply developments elsewhere are being controlled. This emerges from the continued belief in the scarcity of oil to which we referred earlier in the chapter. Although, as shown in the preceding paragraphs, the idea of oil scarcity is not justified by the facts of the oil supply/demand situation, it is nevertheless one of the main elements in the evolution of many countries' policies towards oil. And given such a belief, then it seems to be a more appropriate policy for such countries *not* to produce oil than to produce it; for one or both of two separate but related reasons. In the first place, if oil is going to be scarce in the future then it is going to appreciate in value, so that it is better to leave the oil in the ground, where it will gradually become worth more and more, than it is to produce and sell it. Such an argument justifies limitations on the rates of production in non-O.P.E.C. as in O.P.E.C. countries. Second, if oil is going to become scarce so that it will be expensive to buy from abroad, then it is better for any country which has reserves to hold on to them with a view to avoiding the economic and political consequences of future dependence on imports. This also justifies limiting current rates of production but, in doing so, it also reduces the chances of finding new oil resources, for discoveries of new reserves are, in large part, a function of the continued production of oil. In other words belief in scarcity must lead to scarcity, so that the prediction is self-justifying. Control over oil supply is, in the first place, a cause of scarcity and, in the second place, it is also a result of the scarcity which ensues.

Within a generally controlled oil supply system which thus now seems likely to continue there will also be regulated supply patterns. The O.P.E.C. prorationing system will, of course, be the prime example of this but, in addition, one must also expect the severe regulation of oil production and transportation outside the ranks of the O.P.E.C. producers. Thus, domestic oil will always be given preference over imported oil with a resultant continuing, and perhaps even a rapid, decline in world trade in oil (with consequential adverse results from this for the oil tanker trade). Indeed, it now seems likely that world trade in oil has already peaked. In the future we shall certainly see a relative, even if not an absolute, decline in the importance of internationally traded oil to the total amount of oil which is used, while the essentially international organization of the oil industry will be replaced by a regional structure for the industry, aspects of which will be considered later in the chapter.

Before moving on to that final theme in this analysis of the response to the challenge of O.P.E.C., it is necessary first to look at two further elements of regulation of the oil industry which appear likely. First, alternative imported energies will deliberately be given a place in evolving

267

energy plans so that oil will not in the future be the only sort of energy imported, as it has been in the recent past for most countries. Coal from overseas will be sought in increasing quantities. This is a development which all the major oil companies have already recognized as potentially important, so that they are diversifying into coal production and transportation in countries like Australia, Indonesia, South Africa, Colombia and, of course, the United States and Canada, where the resources can be developed at low cost. Imported liquefied natural gas is also seen as an alternative to imported oil, with Japan in particular already taking large amounts from South East Asia and planning further to diversity its energy imports in this way.

Second, even the output of domestic oil seems likely to be regulated in order to ensure that alternative domestic energy sources are marketable to the levels which governments consider to be appropriate. Such a control over indigenous oil production may, indeed, often be necessary. In the case of the United Kingdom, for example, oil production will have to be restricted in order to enable indigenous production of coal to be maintained, and in order to ensure that there is space in the energy market overall for the planned production of other energy sources.

This best guess as to the future development of the oil industry outside the communist world, in the light of the post-1973 changes to which it has been subjected, is not, of course, a matter of certainty. There are, indeed, a series of major uncertainties which analysts trying to predict the way the world of oil power and the organization of the industry may go in the next few years must face. Currently, however, the major uncertainties all seem to be on the supply side of the question. The likelihood of a radically different demand component from that suggested earlier in the chapter, without prior changes in supply and price, seems to be relatively small. There are, however, several aspects of the development of the oil supply situation which present important uncertainties, in the context of a November 1985 evaluation.

First, we cannot be entirely certain that the solidarity of O.P.E.C. will be maintained. In a situation in which the need for O.P.E.C. oil continues to fall there may well be attempts by one or other of the member countries to increase its share of the total amount of O.P.E.C. oil which it markets. Iraq, for example, considered its level of production prior to its war with Iran to be too low relative to that of other members of the organization. Iraq, moreover, has been trying to develop an economy in which there is a rapidly increasing need for government expenditure and for imports. It can only satisfy these requirements by selling more oil, or more expensive oil. Moreover, as indicated previously, it does have a capacity to expand production and thus, as a result of this combination of circumstances, may try to corner a somewhat larger share of the available market than the other O.P.E.C. members consider appropriate. The war with Iran has eliminated this possibility in the short term, but post-war reconstruction needs will eventually enhance Iraq's needs for more oil revenues. The result of action to corner the market, the possibility of which is by no means limited to Iraq, would place at risk the ability of O.P.E.C. as an organization to survive, and so open up the possibility of a dramatic decline in the price of oil to $18, or even to $12 per barrel. Such an oil price collapse would, of course, influence the evolution of the demand for the commodity by making it relatively more attractive compared with alternatives.

Second, there is a possibility that the United States could so increase its own production of oil, gas and other sorts of energy as to permanently restrict its demand for O.P.E.C. oil. This cannot be entirely discounted, especially if a politico-economic agreement between the United States and Canada not only makes it easier for Alaskan energy supplies to reach the 'lower forty-eight' states via pipelines across Canada, but also makes it possible for increasing, rather than decreasing, amounts of Canadian oil and gas to be exported to the United States. In the 1970s U.S. demand for oil imported from O.P.E.C. countries increased dramatically but, since 1979, U.S. oil imports have fallen away just about as dramatically as they increased over the previous few years. Various options now open to the United States, in respect of alternative energy supplies and in terms of controls over imports, open up the prospect of a continued low degree of dependence on oil imports, especially those from O.P.E.C. countries. This, of course, would have a highly dramatic effect on the international oil situation.

Third, there is also the possibility of really significant growth in oil production potential in other parts of the North American continent, and which additional production could then serve the United States market, via an inherently very secure pipeline system. This relates to the potential of Mexico and of Guatemala. Indeed, Mexico's ability rapidly to increase its production is not in doubt from the point of view of the reserves which have already been discovered. These total over 60,000 million barrels, a quantity of oil which could sustain production well in excess of 300 million tons per year, of which Mexico itself would use under half and so leave an export potential equal to about 50 per cent of the United States' highest level of oil imports in 1979. But the potential oil resource base of Mexico is now put as high as 200,000 million barrels, a level of resources which would put it into the Saudi Arabian class as a potential oil-producing country and so open up the prospect for an oil supply development in the western hemisphere which would upset

O.P.E.C.'s expectations for the demand for its oil in the western hemisphere even in the longer term. Guatemala shares some of the now highly prospective oil regions of parts of the isthmus of Central America with Mexico. Exploration started there in 1979, under new legislative conditions which have recently been established, so that Guatemala, with only a very small internal market, may be added to the ranks of the oil-exporting nations. There is a likelihood that it would achieve this status in the context of a special relationship with the United States, as a continuation of the existing situation between the two countries.

On the other side of the world China also promises the possibility of large-scale oil production. The expansion of its oil industry in the later 1970s was quite dramatic, with annual rates of increase of the order of 20 per cent in production. These have already converted China from an oil importer (it was dependent on the Soviet Union for oil until difficulties arose between the two countries in the 1960s) into a net oil exporter. This is a development which has been of particular interest to nearby Japan, to which China already sells some oil. China is currently somewhat conservatively estimated as likely to have at least 70,000 million barrels of oil resources. Some of the more accessible of these reserves will be exploited in the relatively near future under joint venture arrangements with western oil companies, agreement on the first of which was reached in 1981. Implementation of many of these agreements is now well under way – and some discoveries of oil and gas have already been made in various parts of China's large offshore areas of hydrocarbons potential. In the context of these agreements a production level of up to 200 million tons per year by the end of the century now seems to be a reasonable proposition. As the demand for oil in China itself is likely to be kept restrained by government policies, much of the growth in output (from today's level of about 100 million tons) would become available for export. The prospect of China becoming such an important new supply point for crude oil in the Far East adds an obvious element of uncertainty to the medium-term outlook for international oil supply patterns, and for oil prices.

There are other possible large-scale developments in oil supply in the relatively near-term future, but these are rather more and, in some cases, much more, speculative. Thus, the uncertainty they add to the supply/demand equation for international oil is at the moment much less important. Nevertheless, under the impact of continuing high international oil prices, and with the impact of these still enhanced by continued (even if decreasingly justified) fears over the security of supplies, most countries of the world have initiated or increased their efforts to find indigenous supplies. Many have been entirely unsuccessful but many others have had modest success and a few – such as Malaysia and the Sudan, for example – seem to have hit the jackpot. The cumulative effects of all the new and potential producers on the outlook for O.P.E.C. oil exports' prospects are important and there is, moreover, always the possibility that one or more of them may, like Mexico, find sufficient resources to enable exports to be built up, and in the process also help to eliminate the still generally expected revival of international demand for O.P.E.C. oil by the late 1980s or early 1990s. Major new finds in Western Europe would have such an effect and, of course, if the new oil were found in such a location than there would not be any financial and technological constraints on its development. Elsewhere in the world – notably in the Third World – this might not be true and here the development of large-scale resources will depend upon the achievement of appropriate mechanisms for making exploitation of the resources possible – in a situation in which, as we stressed in Chapter 7, the major international oil companies are not generally welcome in their entrepreneurial guise. Mechanisms, whereby the expertise of such companies can be used in an arrangement which is de-politicized, are being evolved under the auspices of the World Bank and other international and regional economic organizations. These are thus opening up the opportunity for the further expansion of oil supplies from the Third World – including such important countries as India and Brazil, both of which are moving towards possible self-sufficiency in oil. The degree to which this may produce instability in the international oil system would then depend, in large part, on the ability and willingness of O.P.E.C. to find a place for the new producers in the international market, or as members of the organization. This is certainly a problem which O.P.E.C. will have to face more than once in the next decade, especially as its members come under pressure from the World Bank and the other international economic organizations to help, financially at least, with the efforts to exploit the oil resources of Third World countries.

Thus, it is clear that there is a series of oil supply possibilities which, in the context of a very limited rate of growth in oil demand, opens up the question of the degree to which O.P.E.C. will be able to continue to influence the international oil situation. However, because there are important constraints on the speed and the degree with which O.P.E.C. can take positive action to control the market it may, indeed, have to be supported by other parties, particularly by those important energy-using nations which are investing large sums of capital to develop indigenous high-cost energy resources. Such countries (for example,

France with its massive investments in nuclear power) will wish to protect these developments by helping to ensure that the price of oil does not fall to a level at which the investments become uneconomic, and so leave the countries concerned with burdens to bear, rather than with alternative energy supplies to enjoy. Indeed, the International Energy Agency once agreed on the principle of a floor price for oil as a means of encouraging investment in higher-cost energy. In practice, the floor price of $7 per barrel (agreed in 1976) is much too low to be effective, but at least it shows that the principle of fixing a minimum energy price in the western industrial nations has been established. More recently, negotiations were undertaken in Canada between the government and companies involved in developing the Athabasca tar sands with a view to ensuring a market for this so-called unconventional and high-cost oil. As oil from the tar sands needs to be priced at about $25 per barrel (in 1980 $ terms) to be adequately profitable, agreement to such protection by the Canadian government meant that the government was also obliged to pursue policies that try to ensure that the price of international oil is kept at around this level: basically, that means support for O.P.E.C. in its effort to maintain present oil price levels. Similarly, the high tax rates imposed on indigenous oil production by countries like the U.K. and Norway also depend upon the maintenance of high oil prices if the taxes are not to undermine the validity of the investment in relatively high-cost off-shore oil developments.

Finally, the United States government effectively agreed to underwrite the costs of a number of oil-from-coal plants, the competitiveness of which also depends on oil continuing to sell at about $30 per barrel (again in 1980 $ terms). Needless to say, the oil companies also involved in the development of such alternative supplies of energy would have their risks minimized, and their opportunities for earning profits increased, if oil prices were to remain high. And in as far as the companies still exercise influence over the major oil-exporting countries – both through their managerial and technical agreements with them and through their control over refining, transport and distribution systems – they are also in a position to try to influence the price of oil and to help to keep it high. They are, of course, helped in this objective by the control they exercise over the speed of the development of alternatives, most of which are dependent on company expertise.

In other words, many other entities apart from the member nations of O.P.E.C. already see their futures as tied up with the O.P.E.C. interest in a continuing high price for oil. The more that are so involved the more likely it is that their collective hopes will be satisfied. The wealthy industrial countries expect that they will be able to live with the economic consequences of continued high-price energy. They also think that it will be possible to avoid any additional serious political consequences arising from the power which the maintenance of a high oil price would give to the major oil-producing and exporting countries. Optimistically, it is thought that the latter can eventually be absorbed into the western industrial countries' system and, after a necessary period of readjustment, that there will be a return to the 'mixture as before' in the organization and functioning of the system.

The main omission from this sort of argument relates to the lack of consideration for the world's poor oil-importing countries in the evaluation. As shown in Chapter 7 these countries have suffered severe economic penalties from the changed world oil situation and from the consequences of that change for the level of economic activities in the industrial countries. With a few exceptions the situation of the poor countries of the world continues to deteriorate and some are on the edge of national bankruptcy as a result of the high costs of imported oil – and of other goods – and from the lack of demand for their primary products' exports in a western world in severe recession. The continuing failure to tackle the problems of these countries and, in particular, the failure to do something about their burden of high oil prices, introduces the greatest degree of uncertainty into the outlook for the system as a whole. Stabilizing world oil prices at their present, let alone higher, levels and simply formalizing the oil power situation as it happens to have evolved over the last decade seems almost the least likely way of ensuring stability in the system overall. The changes in the world of oil power have pushed the western economic and political system into the beginnings of a New International Economic Order. The oil exporters' oligopolistic behaviour seems certain to produce a significant effect on other groups of nations hitherto excluded from a 'say' in running the system. Thus the development we have seen in the last few years in the world of the oil-producing nations could, on the one hand, be but the precursor to wider and even more fundamental changes in the organization of trade in the commodities which are required for development and for improving living standards. On the other hand, the increasing economic problems for the poor oil-importing countries may eventually drag the whole western system down from the levels of development which have been achieved in the post-1945 period, and so open up a whole new set of politico-economic issues for the last twenty years of the twentieth century. Developments in the first five years of this period (1980–85), with record unemployment in the industrialized countries and massive indebtedness amongst the developing countries, indicate that a pessimistic outlook may be more appropriate in the continued absence of any effective moves towards new international agreements which meet some of the needs of all constituent parts of the western economic system.

The oil industry continues to make news daily. Indeed, developments in oil affairs seem increasingly to impinge on the whole range of issues that constitute the complexity of the modern world system. Particular news items about oil, however, can be interpreted only in the light of the increasingly complex world oil power structure which this book has attempted to describe and explain. The traumatic events in the oil world since the first edition was written indicate that the interrelationships of the several sets of 'actors' involved have become more, rather than less, complex. They have also produced a realignment of forces which made the world of oil in the second half of the 1970s significantly different from the one to which we became used between 1945 and 1973. The outlook for the rest of the 1980s and the 1990s seems likely to be as different again and it is difficult to forecast because the extrapolation of previous experience is no longer valid. A new structure is a possibility.

Earlier in this chapter we have referred to the de-internationalization of the global oil industry and to the dramatic decline in the use of oil as part of the response to O.P.E.C.'s challenge. As these processes seem likely to gather strength over the next few years and so come to constitute important elements in the evolving world of oil power in the 1990s, they are worthy of our closer attention in the final part of the chapter.

We have, indeed, argued that there will be a very slow, or perhaps even a negative, growth in the non-communist world's use of oil. Oil is, indeed, rapidly becoming the energy of last resort in all markets where the O.P.E.C.-controlled price has had to be accepted or has been voluntarily introduced (as in non-O.P.E.C. oil-producing countries in which governments have not kept down the prices of locally produced oil). This process has already severely undermined the growth of international oil exports and is thus leading to the reversal of the earlier emphatic development of the international nature of the oil industry.

Concurrent with this demand-side change in the outlook for oil, but achieving importance even more slowly, is the supply-side response to the twelvefold rise since 1970 in the real price of internationally traded crude oil (see Figure 1). This is taking the form of an increasingly geographically dispersed pattern of oil exploration and exploitation. The slowness of this change is not only a function of the long lead times involved in establishing new oil production capacity, but is also a result of the reorganization of the industry which is needed to enable it to adjust its exploration and production efforts away from its hitherto near-exclusive concern with the low-cost, familiar and relatively low-risk opportunities for oil exploitation in the O.P.E.C. countries and North America. Nevertheless, the process is under way, it is intensifying, and it is gradually producing a more diversified geography of oil production. This means that a steadily increasing number of countries are, first, reducing their import requirements, and then, in many cases, achieving the prospect for self-sufficiency in oil. Though this is being done at costs which are often much higher than the resource costs involved in the production of their oil requirements in the O.P.E.C. countries, the costs involved are still lower than the expenditures incurred to acquire oil imports from O.P.E.C. countries, at the international prices which have been established over the last decade by the powerful combination of politico-economic forces involved.

There is thus an inexorable move towards a diffuse and dispersed pattern of oil production which, if it continues, will produce a restructured non-communist world oil industry in which four components will be of particular importance. First, exports of oil from the traditional exporting countries will become the residual, rather than the central, element in the system. Second, non-Middle East members of O.P.E.C. could find their best interests served by close association with, and/or membership of, regional organizations devoted to the establishment of high levels of energy self-sufficiency. Third, there will be increasingly intense competition between those countries with large potential export surpluses of energy, notably oil and natural gas, which are not linked into increasingly energy self-sufficient regions of the non-communist world. This applies especially to the Soviet Union with its economic need to export energy to the west and to the major oil exporters of the Gulf. The latter will be increasingly obliged to use their oil/gas production as low cost inputs to energy-intensive industry as a way of maintaining export earnings. Fourth, the world-wide infrastructure and organization of the industry will become even more complex, in terms of physical elements such as transport and refining/processing facilities and the structure of those oil companies which decide to try to continue to operate internationally as oil companies, rather than diversifying into other activities or retreating into North American operations only.

One likely result of the impact of the factors outlined above could be the emergence of a number of wholly or largely oil-self-sufficient regions, within each of which an increasingly integrated set of linkages will be developed between producing, refining, transhipping and consuming areas. Concurrently, the external connections of each region in respect of oil would become of decreasing importance, and eventually even of an intermittent nature. World trade in oil, and the international organization of the oil industry with which we have become so familiar, would be replaced by intra-regional trade and a regional pattern of industry organization, respectively. It is not yet possible to define, with any degree of certainty and precision, the shape of this possible new international system of world oil regions but there are a number of pointers which hint at the shape of the developments to come. These are discussed below and illustrated in Map 12.

An East/South East Asia region is perhaps the clearest possible potential development. On the producing side Indonesia, Brunei, Malaysia and Thailand are already significant contributors, or are poised to contribute more emphatically, to the regional supply. Further geographical diversification of production can also be anticipated as a result of expanding exploration and development efforts in a region where oil and gas prospects are not only extensive, but also good. Organizationally, Singapore provides the fulcrum for the region and the already noticeable strengthening presence there of many oil companies can be expected to intensify. The region's relatively rapid rising demand for energy (a function of Japan's continued economic growth, and the more recent industrialization of a number of other countries in the region such as South Korea and Taiwan) provides the incentive for enhanced levels of production, particularly for Indonesia, with its massive potential for natural gas production and an expanded oil-producing industry. Guaranteed outlets in the region for its energy exports eventually seem likely to outweigh the future importance of its membership of O.P.E.C., so that it would be prepared to link its future to the energy interests of the region. For Japan, both guaranteed energy availability from relatively nearby producers, and the opportunities provided for its exports in the region, indicate a possible resurgence of Japan's earlier idea of a 'greater co-prosperity sphere' for this part of the world.

The next most likely region is one in Middle America, the focus of which lies in the southern Caribbean in an area of the hitherto divisive rivalries of Venezuela and Mexico. The two countries have, however, already jointly accepted responsibilities for supplying oil to their poor neighbours on favourable terms, whilst their large long-term future export potential of conventional and/or unconventional oil provides a secure resource base for the whole of the region, including the potentially very large needs of Brazil as well as other smaller countries. These prospects, coupled with the increasing chances of tying the U.S. into a western hemisphere energy supply system, suggest a net advantage for a regional

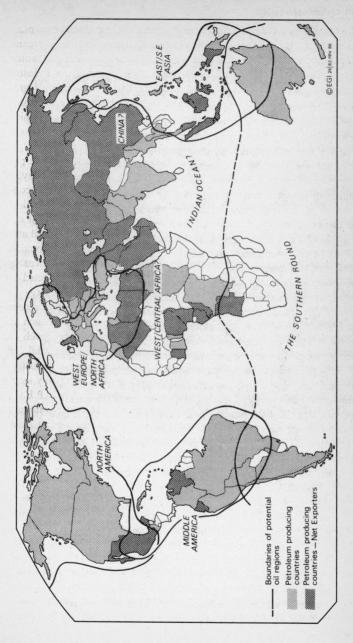

Map 12. Possible future world oil regions?

commitment, in place of continued O.P.E.C. membership, for Venezuela and Ecuador, in cooperation with Mexico and other potentially large oil (and gas) producers in the region. The region's organizational fulcrum is less evident than in the case of Singapore, but the pre-existing oil refining, transhipping, trading and organizational functions of the Netherlands Antilles, together with their relative political stability, did, until recently, suggest one possibility, particularly as they enjoyed a status of neutrality between the interests of Venezuela, Mexico and the United States. Unfortunately, their potential in this respect has now been undermined by the 1985 decisions of Exxon and Shell to abandon their large-scale operations in Aruba and Curaçao respectively, on grounds of adverse economic conditions and prospects. The oil interests of the islands can only be saved through Venezuelan intervention (Venezuela's state oil company has leased the former Shell facilities on Curaçao for an initial three years) so that the 'neutrality' of a Netherlands Antilles base as Middle America's organizational fulcrum for the region's oil industry is no longer extant.

A prospective world oil region without the same degree of close geographical cohesion is that of the 'Southern Round'. This comprises the countries of the southernmost parts of the continent of Latin America and Africa, together with Australasia. Oil and gas prospects in Argentina and on its extensive continental shelf, plus those of Australia and New Zealand provide a basis for the region's self-sufficiency for many decades into the future, whilst common interests, arising from global geo-political considerations, could provide an increasingly powerful motivation for cooperation over energy supplies. The elimination of the need for imports from other parts of the world would thus undermine the impact of the embargo on oil supplies to some parts of this region by the member countries of O.P.E.C. (this was discussed in Chapter 8), so that the embargo would then become even less significant than in the current situation. The expansion of the region of the 'Southern Round' to take in the fringes of Antarctica, with its promise of hydrocarbons potential, would extend the scope of the region both geographically and temporally. Indeed, such a development could open up supply prospects and the need for markets outside the region by the turn of the century.

A fourth region in prospect, though one which currently has little internal cohesion, comprises Western Europe and the Mediterranean basin. Here, the large, and potentially much larger, oil- and gas-producing countries of the North Sea basin and adjoining north-western European offshore areas, plus the easily linked-in suppliers on the southern side of the Mediterranean Sea, provide the reserves and the production potential which could eliminate the dependence of this most intensive energy-using region of the world on supplies from other parts of the world. Refusals to license potential producing areas, hesitancy in allowing oil and gas which is discovered to be produced, and the propensity of many of the countries concerned to tax the oil companies into lethargy as far as their investment plans for the region are concerned, combine to produce a case-study on how not to approach the development of resources. Nevertheless, there are fundamentally more powerful forces which could come into play to create a potential for cohesion over energy developments. These include, first, the growing impact of an increasingly hostile external world in general, and threats from the major powers, in particular, to the security and/or welfare of the region: and, second, the recognition by countries as diverse in their current interests and policies as Algeria, Libya, the United Kingdom and Norway that guaranteed markets in Western Europe for their oil and gas are preferable to the uncertain future which faces their indigenous oil and gas industries as a result of the changing international energy/oil situation.

Changes in policies consequent upon the influence of these factors could provide a region potentially self-sufficient in oil and gas production. Moreover, most of the refining, transport and distribution infrastructure required to get the quantities needed to the markets is already in place, or it can be easily and quickly developed. External relationships in respect of the future energy import needs of the Western European/Mediterranean region, such as they will then need to be, will be to the east. From there the Soviet Union, on the one hand, and countries of the Gulf on the other, will have to compete for the very limited markets which are likely to be available.

The oil regions discussed above by no means cover the world. Other regions, as shown on Map 12, are possible – including West/Central African and circum-Indian Ocean regions – but there are still insufficient indications of their development to justify even initial speculation about them. China is also excluded. It could be the producing centre of an East Asian region with links to its neighbours to the south and, even more important, to Japan; but a more likely prospect is its integration into the East/South East Asian region already described. On the other side of the Pacific, a region comprising Canada plus the United States is an alternative to the latter's involvement with the Middle America region, providing Canada continues to pursue its recently developed less isolationist oil/energy policy.

This hypothesis of a world of largely independent oil regions has, of course, important implications for the main actors in the current global oil system. This, of course, involves consideration of the position of the Soviet Union, the oil-producing states of the Gulf, and the international companies. The Soviet Union appears to be wealthy enough in its potential resources to be able to sustain the future oil and gas needs of the eastern block for the rest of the century. Given positive co-existence with the United States, leading to technological and financial help for the more effective and rapid development of the Soviet oil and gas potential, it would also be able to expand its present commitment to supply energy to Western Europe. As already noted, however, a cohesive Western Europe/Mediterranean region will have limited needs of oil and gas imports. In this situation the Soviet Union would have to compete against the other contemporary major oil regions of the world that we hypothesize as being excluded from a world of oil regions, viz. the Gulf, the heartland of the international oil system over the last thirty years.

One could argue that the major exporting countries of the Gulf have themselves created the conditions which necessarily exclude them from the future world of oil. These are the consequences of policies which have pushed the price of their oil exports so far above the long-term supply price of the commodity as to make their production decreasingly relevant in a world which, with a much lower growth rate in oil use, has alternative oil supply options open to it. This is especially true when the economic element in the argument is coupled with the fears for the security of the oil supply potential of the region as a result of issues such as the Arab/Israeli dispute, the conflicts and potential conflicts between contiguous oil-rich countries of the region, and the rise of Islamic militancy.

The significance of the potential for a possible reshaping of the world oil industry in the relatively near future along the lines indicated above will not, however, be lost on the Gulf producers. They may thus be expected to try to contain the developments suggested. They could, for example, use the power they are able to exercise over nations currently dependent on oil imports from the Gulf to persuade them not to seek regional agreements for self-sufficiency in oil. Or, they might attempt to 'buy' their way into the regionalizing system, with Western Europe as their most likely regional objective in this respect. Failure would diminish the importance of the Gulf oil-exporting countries not only in world oil terms, but also more generally in world geo-political terms. One way of avoiding this would be if they decided that their interests were likely to be best served by a close alliance with the U.S.S.R., in order to achieve an agreement on a market-sharing arrangement for the limited oil and gas export opportunities in a world of near self-sufficient oil regions. This would be a highly paradoxical development, given current fears in some circles over Soviet plans to take control over the Gulf region in order either to ensure its own oil supplies, or to deny oil to the west. It would certainly be a geopolitical change with immense implications and the possibility, in itself, may suggest to the western industrialized world that its oil imports from the Gulf ought to be maintained at levels higher than those which would remain, were the process of the re-adjustment of the world of oil into wholly or largely self-sufficient regions to proceed unhindered, on economic and geopolitical grounds.

Finally, it is clear that the international oil companies would also need to rethink their global strategies for continued profitable operation if the world oil system were to evolve as we have hypothesized. They would need to reshape their own organizational and control mechanisms in order to reflect the newly emerging regional pattern of the late twentieth-century oil industry. Some of the companies seem likely to be more flexible than others in respect of these needs, and thus be able to retain an element of managerial and technological internationality in the context of the regionalized oil system. Some of the companies, on the other hand, may decide to pursue a policy which restricts their oil and other activities to the United States, in order to ensure what profits they can in an organizationally much less demanding 'fortress America'. However, the impact there of the long held belief in, and the likely enhanced efficacy of, government controls over oil companies (the Reagan administration excepted), as well as the absence in the United States of inherently low-cost oil and gas resources, may make it more difficult for oil companies to continue to earn a respectable living in the U.S., compared with the greater, though more risky, chances for profitable operation in the more challenging world outside, where, as shown previously in this chapter, there are still many opportunities for oil and gas developments on a large scale. The prospects for these would be much enhanced by the evolution of a world of oil regions.

This last chapter has attempted to specify the responses to the initial successes of O.P.E.C. and the implications of those responses in terms of how the world of oil power might evolve over the remaining years of the twentieth century. The dynamics of the industry, however, are now so great that even the relatively short period between writing the text and its publication may prove the author wrong in respect of the short-term outlook. If this has indeed proved to be the case then he hopes, instead, that in his attempt to understand the situation he will have created an awareness of the broad set of fundamental issues involved in the new world of oil, and so enable each reader to undertake for him or herself a reinterpretation of the position and the outlook in the light of the most recent events.

Postscript

In the last paragraph of Chapter 10 the author noted that the 'dynamics of the [oil] industry . . . are now so great that even the relatively short period between writing the text and its publication may prove [the author] wrong in respect of the short-term outlook'. And so it has proved. In November 1985, when the completed manuscript was delivered, the international price of oil was just under $30 per barrel. It is now less than $13 and it is not generally expected to recover to more than $20 – at most.

The possibility of 'a dramatic decline in the price of oil to $18, or even to $12 per barrel', was suggested in Chapter 10 (p. 269), but this was thought to be less likely than a continuing steady decline in the oil price in spite of 'the question of the degree to which O.P.E.C. will be able to continue to influence the international oil situation' (p. 271). The author's conclusion in this respect was reached because he thought that many other parties ('particularly important energy-using nations which are investing large sums of capital to develop high-cost indigenous energy resources' – p. 271) would help to ensure that the oil price did not collapse. This has turned out to be incorrect: no country or other institution concerned with oil has yet taken any significant action to rescue O.P.E.C. from the rapid decline in oil prices which started in mid-January 1986. This is partly because so many other parties – notably importing countries and consumers in most parts of the world – have welcomed the lower oil prices, and partly because even those parties which did have an interest in keeping prices up saw that the members of O.P.E.C. were themselves failing to take action which would have this effect. Indeed, in the face of the very weak and potentially over-supplied oil market which had emerged by 1985 (as described in pages 249–64), first, individual members of O.P.E.C. and, then, O.P.E.C. as an organization, took action in the second half of the year which made a bad situation even worse. Most members were, in one way or another, already trying to sell more oil than the quota they had accepted at the O.P.E.C. meeting in January 1985, but the impact of their action was disguised for a time by Saudi Arabia's willingness to allow its exports to continue to fall so that the market was not swamped with too much oil. By the summer of 1985 Saudi Arabian oil production was down to only a little more than 2 million barrels per day (100 million tons per year). At this stage the country's rulers decided that the situation had deteriorated far enough and that their production and exports must be increased. Saudi Arabia warned its fellow O.P.E.C. members of its intentions, but to little or no avail in terms of the willingness of other O.P.E.C. members to reduce their levels of production. After September additional supplies of Saudi oil began to reach world markets, under the terms of new purchasing agreements which the country had made with the major oil companies and a number of other oil traders. As these were attractive to the companies (in that they guaranteed profits at the refining stage by indexing the crude oil price to the market value of the products made from it – the so-called 'net-back' pricing arrangements), there were no immediate difficulties in Saudi Arabia achieving its objectives of more production and exports. Thus, by the end of 1985 its production had doubled and its exports more than doubled. As a result, the potential world over-supply of oil, as earlier in the year, became an actual over-supply, though this was masked for a time by the normal winter stock-building programme and an early, very cold 'snap' in Western European and North American weather. Thereby, the demand for oil was briefly stimulated above expected levels for the time of the year.

This was followed by a period of unusually mild weather, a development which in itself was enough to cause the market to weaken. It also, however, coincided with an important mid-December decision by O.P.E.C. ministers – at one of their regular meetings. They decided to terminate their hitherto agreed (though not always implemented) policy of restricting their countries' total production of oil to the volume which was estimated to be needed to equate world supply with world demand – after taking non-O.P.E.C. production into account. As shown above (pp. 250–51), this policy had already had the effect of halving O.P.E.C.'s contribution to world oil supplies. This was a situation that the oil ministers of the O.P.E.C. countries found unacceptable and they now declared instead that the new objective of the organization would be to secure what was described as its members' 'rightful and reasonable' share of the world oil market. What the ministers considered to be 'right and reasonable' was not defined, but the declared new aim of O.P.E.C. clearly meant that more oil would be brought on to the world market as the collective decision by O.P.E.C. gave the signal for all member countries formally to abandon their adherence to the quotas which they had hitherto accepted – in principle, if not entirely in practice.

There was, of course, no possibility that world demand for oil would suddenly increase in the short term, so to absorb O.P.E.C.'s additional production and exports. Thus, the only remaining hope for oil price maintenance lay in the ability and willingness of non-O.P.E.C. producers to reduce their output and so re-balance overall supply and demand. Indeed, O.P.E.C. appeared to believe that its decision to seek a larger market share and thus threaten existing prices would so frighten other producers that the latter would more or less immediately announce their compliance with O.P.E.C.'s expectations. Britain and Norway, the two major North Sea producers responsible by the end of 1985 for production at an annual rate of more than 150 million tons, were specifically selected for O.P.E.C.'s attention and pressure. This was partly because they were more appropriate targets politically than the world's two major producers, the United States and the Soviet Union (which together produced over 1,150 million tons in 1985); partly because their economies were seen as being dependent on the maintenance of high oil prices; and partly because the North Sea oil's development was perceived by the members of O.P.E.C. as the main new element in the world oil supply pattern which had undermined O.P.E.C.'s control of the market – an aspect of the changed international supply situation since 1973 to which much attention was given in Chapters 5 and 10. O.P.E.C. appeared also to expect that the main non-O.P.E.C. oil-exporting countries in the Third World – notably Mexico, Egypt, Oman, Malaysia and Angola – would more or less automatically fall into line with the new policy requirements and so willingly and quickly reduce their exports.

O.P.E.C.'s pressures and expectations were not realized in any of these respects so that the average price of internationally traded oil fell from about $28 in mid-January 1986 to just over $15 per barrel by mid-March. It then temporarily stabilized, prior to the O.P.E.C. ministerial meeting in March, in the expectation that some action would be taken by O.P.E.C. to reintroduce production quotas and controls adequate enough to secure a return to a price of around $20 per barrel – a price which by that time had become very generally accepted as the highest that could be sustained in the unbalanced potential supply/demand situation. The meeting of O.P.E.C. failed, however, to produce the appropriate action, and the price of oil fell further to an average which was only a little above $13 per barrel. The 'spot' prices of certain internationally traded crudes (such as North Sea Brent crude and Dubai crude), and of the United States' marker crude (West Texas Intermediate grade) dropped to even lower levels – down to $10 per barrel – by the first week of April 1986 and have only recovered slightly by early May as this postscript was completed.

Mid-1986 is thus a time of more uncertainty than usual and of greater complexity than ever in the international oil industry and in its impact on international affairs – both economic and political: and there are still some months to go before this edition of *Oil and World Power* will be published. During this period there will be continuing discussions and negotiations between, and actions and reactions by, the very large number of interested parties in the international oil situation and prospects. The following brief evaluation of the early May 1986 situation seeks to do no more than to recapitulate the essential geo-economic and the geopolitical issues which are involved in the light of the late 1985/early 1986 changes in the world of oil power. The issues presented will still be of the essence when the book is published – though some of the details may have changed – and will thus provide a specific context (along with the more general context provided by the book as a whole) within which to interpret the continuing sequence of events concerning international oil.

It would be a gross misjudgement of the impact of O.P.E.C. over the past fifteen years to suggest that its inability to keep up the price of oil means a return to the *status quo ante* of the first oil-price shock of 1974–9 (see Figure 7) – in either world economic outlook terms, or in respect of international geopolitical questions. The changes in the Western world's economic situation arising from the oil-price changes of the 1970s have run deep. This is so not only in respect of their impact on economic growth rates and on employment in the industrialized world, but also in limiting the aspirations of many countries in the developing world to achieve higher standards of welfare for their inhabitants. Even in the still rather unlikely circumstances that oil prices will fall to – and remain at – about $10 per barrel for some time (that is, at about the same price – in real terms – as they were in the early 1970s, as shown in case Cm in Figure 7), it would certainly be almost two decades on from the initial disruption of the international oil market in 1973 before things were back to 'normal'. With prices more likely to hold at or about the level following the first oil-price shock (as shown in case Bm in Figure 7) the economic outlook for the Western system is less favourable, so extending the period over which the problems caused by O.P.E.C.'s success will remain noticeable, painful and debilitating. Thus the changes – albeit temporary – of such significant proportions in the price of such a universally required commodity since 1973 can be interpreted as the single most important factor which turned the Western economic system

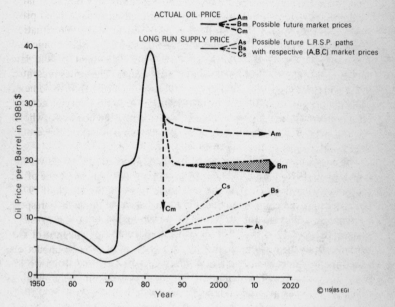

Figure 7. The long-run supply price and the market price(s) of oil, 1950–2020

Am Slow downward drift of oil price under conditions of institutionalized intervention in the market, leading to continued decline in the demand for oil; and continued instability in the market.

Bm Near-future (1986) price cut of 25–30 per cent, so stimulating demand but with continued propensity for investment in oil exploration and exploitation. Thereafter, relatively narrow long-term price range under conditions of sustainable supply/demand relationships.

Cm Under price collapse conditions the actual oil price returns to a level close to the L.R.S.P. Demand stimulated with consequential steeper slope to the L.R.S.P. curve, so creating a potential for relatively near-future renewed upward price pressures and price fluctuations.

sour after twenty-five years of expansion and which still casts a shadow of doubt over the prospects after more than a decade of difficulty. Even viewed in these narrow geo-economic terms, the impact of O.P.E.C. on the West has been traumatic.

The significance of O.P.E.C. is, however, even deeper and more fundamental than this in spite of the fact that as an organization it was, for the first decade of its existence, essentially ignored both by the international oil companies and by the industrialized world's governments. This was basically because there was no expectation by these powerful entities that it could, or would, have any relevance to questions of the supply and price of oil. Thus, the companies treated with O.P.E.C. in the 1960s only in respect of technical questions, and they certainly did not visualize it as any kind of a contributor to the strategy of international oil – even when, after 1968, they decided to 'use' the pressures of O.P.E.C. as the scapegoat for justifying the increased prices they had to charge for oil. They needed such higher prices to finance the expansion of the industry, but they thought themselves powerful enough to keep control over the system in spite of the way in which they used O.P.E.C. at that time.

The companies were thus shocked by the O.P.E.C.-imposed oil price increases in 1973–4 and 1979–80 but, between the two price shocks, there were developments in the relations between the oil-exporting countries and the companies that were of even more fundamental importance in changing the structure of power in the international oil system. Between 1974 and 1979 the companies were simply dispossessed of the bulk of their assets as, one by one, O.P.E.C.'s members nationalized the oil concessions which the companies had secured from them over many decades. The compensation they received was, in general, derisory, as it was generally related to the much 'written-down' value of the companies' assets – a record of value, moreover, which was related essentially to tax-minimization objectives rather than to the worth of the oil they had discovered in the concession areas.

Dispossession was thus traumatic for the oil companies which had hitherto been the most powerful set of interests in the Western world's system. It was, however, also irreversible as it is impossible to conceive of circumstances in which ownership of the oil reserves and resources could be re-obtained by the companies – except by force of arms. This wholesale nationalization of the oil companies represented a breakthrough in relationships between developing countries and multinational corporations – given that the former achieved a degree of control over the use of their national resources. This had not hitherto been seriously considered likely, except in the largely academic discussions on the achievement of a New International Economic Order. Progress towards the latter had hitherto been very tentative, but given what happened in the world of oil between the international oil companies and the major oil-exporting countries, a firmer base was established from which the developing world could work. It seemed likely that late twentieth-century developments in international economic relationships might eventually be achieved – and so change the shape of the Western economic system.

The changes induced by O.P.E.C. have, indeed, been significant for North/South relations at the intergovernmental level. The failure by the industrialized countries to recognize O.P.E.C. in the first decade of its existence was replaced after 1973 by apprehension over the threat posed by O.P.E.C.'s actions to undermine the Western economic – and political – system. At that time there was, as indicated in Chapter 9, much talk of the need to destroy O.P.E.C. before it destroyed the system. In other words, there was still little or no degree of acceptance in the industrialized world of the 'right' of a group of countries in the Third World to join together in action to secure what they saw to be their own best interests – in the same way as the industrial countries had done over many decades, and especially since 1945 with organizations such as the O.E.C.D. and the E.E.C. Instead, O.P.E.C. had to struggle to secure recognition of the aims which it was pursuing on behalf of its members. It was successful in the short term in this respect and, through the price increases which it won, O.P.E.C. achieved greater immediate financial rewards for its members than any other international economic organization in the history of the Western system.

The seriousness with which O.P.E.C. came to be taken by the industrial countries after the second oil-price shock was indicated in the way in which concern for energy questions in general, and for oil issues in particular, were elevated to a position of prime importance in international discussions. This even extended to the frequent discussion of international oil and energy questions at the regular summit meetings of Western leaders which were held over this period. Reaction against O.P.E.C. even became somewhat frantic by the late 1970s at the time of the second oil-price shock. As a result the International Energy Agency was instructed by its member governments to direct its main efforts to the implementation of policies which sought to minimize the use of O.P.E.C. oil – no matter what the cost. This objective was spelled out clearly in the Organization's 1980 *Annual Review*:

> The 1980s must be a period of major transition towards a minimum oil economy
> ... Necessary adjustments ... must continue to be a focus of energy and economic

planning. Policies of I.E.A. countries need to be reinforced and extended to . . . achieve these ends.

The result of this near-obsession with the dangers of 'reliance on O.P.E.C. oil' was action which, in essence, was contrary to the basic economic philosophy of the industrialized countries in respect of free international trade and the preferred use of lower-cost goods. The energy policies which were advocated deliberately sought to restrict international trade in oil and encouraged, and even required, the use of sources of energy which were inherently more expensive than oil. The fact that the industrialized Western countries should have so acted was a powerful expression of what had hitherto been an unknown concern for the impact on the Western political and economic system of a group of nations in the 'South'. As such it represented a new departure in international economic and geopolitical relationships in the Western system.

The question as to whether the O.E.C.D. countries have by now become willing to accept the validity of O.P.E.C.'s existence, and of the changes which the oil exporters have brought to the Western economic and political system, comes with the dangers for the survival of O.P.E.C. as a result of the weak international oil market and the increasingly ready availability of alternatives to O.P.E.C. oil. O.P.E.C. does, indeed, face great difficulties in saving itself, given the rising pressure of potential supplies of oil and energy in the context of a near-stagnant demand for oil in the non-communist world outside the O.P.E.C. countries themselves (as shown in Table 3 on page 251). However, a number of O.E.C.D. countries, as well as many powerful institutions in the financial and commercial sectors of the industrialized world's economy, are also suffering adverse effects by the sharp and rapid fall in oil prices which has occurred in the first few months of 1986. Thus, it may be argued, O.P.E.C.'s continued existence as an organization with which discussions over oil and energy supplies and prices can take place, as well as the maintenance of the higher economic standards to which the populations of the oil-exporting countries have become used, is now a requirement for the overall long-term well-being of the Western system. O.P.E.C., in other words, is an organization to be protected and cosseted, thereby incorporating into the effective part of the Western system a group of countries which have hitherto had little claim on its wealth and even less claim on the right to participate in its decision-taking. The O.P.E.C. countries are not, however, countries whose progress economically and politically has been in the traditional way of the Western world, i.e. through the achievement of wealth and prestige by means of industrialization. Their acceptance as effective contributors to the system therefore necessitates a change of attitude and approach by the present members of the O.E.C.D. It thus creates problems of acceptability.

There are two other aspects of importance in the enhanced role of the oil-exporting countries in international affairs since 1973. The first relates to the geopolitical, strategic and cultural importance of the Middle East, where most of the member countries of O.P.E.C. – and most of the world's proven reserves of oil – are located. Clearly, the predominantly Judaeo-Christian-based set of rich industrial countries, with its powerful political and strategic interests in the Middle East generally, and in the Gulf in particular (partly as a result of its proximity to the Soviet Union), has thus been confronted, through the changes in the world of oil power, with an additional dimension in evaluating its relationships with the region. In essence, the Western oil companies' loss of ownership of the oil reserves of the region, their loss of control over decisions on their exploitation and the general perception that the West requires access to these resources have necessitated a much more active, responsive and more carefully considered policy towards the Arab oil-exporting countries. In the first instance, as shown in Chapter 9, the needs arising out of the oil crises produced a break in the earlier high level of Western solidarity with Israel. Policies relating to Israel and the Arab states of the region had to become more even-handed. Even more fundamentally, the question of the West's relationship with the Islamic world of the Middle East (and elsewhere, given that two of the other five member countries of O.P.E.C. – Nigeria and Indonesia – are also partly Islamic) has become an issue of importance much sooner than would have been the case had there not been the transfer of wealth and of power to O.P.E.C.'s members as a result of the oil-price shocks.

This aspect of the changed situation is most specifically an issue of great importance to Western Europe – as, in economic as well as political terms, it points to the increasing validity of policies which are based on the French concept of 'the Middle East as an extension of Europe' (see above, pp. 237–40). Thus, preliminary discussions with Turkey over its possible accession to the E.E.C. have already taken place and the issue of its 'membership of Europe' is likely to become a central one in European politics by the end of the 1980s. If and when Turkey is admitted to the E.E.C., then a geographical barrier which was erected in the twentieth century between Europe and the Middle East will have been broken (and earlier geographical ties renewed). Thereafter, further integration between Western Europe and Turkey's eastern neighbours – the oil-exporting countries of the Middle East – would seem likely to become an unstoppable process. In part at least, this would be in recognition of a mutuality of interest arising out of the existence of long-term energy markets in Europe on the one side, and from the long-term availability of low-cost oil and gas on the other.

Finally, the continued existence of, and an important role for, O.P.E.C. in international discussions is important for North–South relations. Though the O.P.E.C.-generated oil-price increases of the 1970s were little short of an economic disaster for the rest of the Third World, there has, nevertheless, remained a relatively high degree of solidarity between the oil-exporting countries and most of the rest of the countries of the South. This has been helped by the fact that many of the oil exporters have extended programmes of economic assistance to the poor countries. Many of these programmes have, indeed, when related to population or per capita income of the donor countries, exceeded the direct help given to the poor countries by most of the industrial nations of the North. Moreover, the success that O.P.E.C. had in partially undermining the power of the rich industrial countries created a great deal of admiration in the rest of the Third World. Thus the demise of O.P.E.C. would, very generally, be regarded as a retrograde development in international political and economic terms; particularly as its demise would certainly be presented – and not without justification – as the result of a deliberate counterattack by the industrialized countries of the North; a counterattack, moreover, which would certainly be interpreted not only as being aimed at O.P.E.C. as such, but also as a means for maintaining the status quo between the industrialized and the developing countries in terms of the division of power in the system. In other words, the continuity of O.P.E.C., as an organization of a number of the world's developing countries, is required if there is not to be a further exacerbation of the already less than satisfactory North–South relationships within the Western system.

O.P.E.C. and the succession of oil crises since the early 1970s have thus changed much more than the price of oil. Indeed, the latter – measured in real terms – has, as shown above (pp. 285–7), now fallen back close to the long-run supply price of the commodity – and, as a result of the easy long-term relationships between oil (and energy) supply and demand, it seems unlikely to recover to very much more than half of its highest-ever value in 1981 of just under $40 per barrel. The longer-lasting components of the existence and work of O.P.E.C. and of the oil crises of the last decade and a half are more generally economic and geopolitical. These effects are more fundamental than the price changes, and they may yet prove to mark a turning point in the history of the

Western system – much more so, indeed, than would have been the case if the oil crises had simply been the harbingers of economies obliged to run on sources of energy other than oil.

The era of oil as the world's main energy source still dates back for less than two generations and there was, as shown above in Chapters 9 and 10, a widely held view in the 1970s and early 1980s that it was on its way out as an important source of energy for the world's economies. Given the now much lower price of the commodity and the now wider recognition of the validity of the arguments of those of us who said that oil was not scarce, the future of the world oil industry now seems likely to be very much longer than its past. Its future role, however, could well be in the context of a Western system which will turn out to have been significantly changed by the politico-economic forces which were released as a result of the initial successes of O.P.E.C. – and the temporary energy supply and pricing crises which it caused in the 1970s and the early 1980s.

May 1986

Suggestions for Further Reading

1. Introduction: The World's Oil Industry

An earlier book by the author, *An Economic Geography of Oil* (Bell, 1963, and revised edition in preparation), analyses the locational patterns of the world oil industry on a function-by-function basis while J. P. Riva, *World Petroleum Resources and Reserves* (Westview Press, 1983) is concerned with the geological basis for the current world petroleum situation. E. T. Penrose's *The Large International Firm in Developing Countries* (Allen & Unwin, 1968) provides an excellent description and interpretation of the economics and organization of the international oil industry in the post-1950 period of expansion. *Essentials of Petroleum* by P. H. Frankel (2nd ed., Cass, 1973) and *The World Petroleum Market* by M. A. Adelman (published for Resources for the Future by the Johns Hopkins U.P., 1972) provide, respectively, an introductory and definitive view on the economics of the industry in its pre-O.P.E.C.-controlled state. *Multinational Oil* by N. H. Jacoby (Macmillan, 1974), *Oil Companies in the International System* by Louis Turner (2nd ed., Allen & Unwin, 1983) and *The Control of Oil* by J. M. Blair (Macmillan, 1977) provide more recent interpretations of the industry's international structure. *The Evolution of O.P.E.C.* by A. L. Danielsen (Harcourt Brace Jovanovich, New York, 1982) provides a comprehensive and rigorous analysis of the world oil market covering both theoretical economic and institutional aspects. Two books by authors whose work as specialized journalists put them closely in touch with the working of the oil industry fill in some of the industry's political and historical backgrounds. These are *Oil Companies and Government* (Faber & Faber, 1967) by J. E. Hartshorn, and *Oil: the Biggest Business* (Eyre & Spottiswoode, 2nd ed., 1975) by C. Tugendhat and A. Hamilton. A. Sampson's *The Seven Sisters* (Hodder & Stoughton, 1975) looks at the world of oil specifically from the way in which it was organized by the seven major international oil companies. H. O'Connor's *World Crisis in Oil* (Elek Books, 1963), M. Tanzer's *The Energy Crisis* (Monthly Review Press, 1974) and R. Engler's *The Brotherhood of Oil* (Chicago, 1977) provide a markedly different interpretation of the history and politics of the industry. The reader seeking understanding not only of the issues, but also of the emotions involved, will want to read both points of view. For an account from the point of view of O.P.E.C., see M. S. Al Otaiba, *O.P.E.C. and the Petroleum Industry* (Croom Helm, 1976), F. Ghadar, *The Evolution of O.P.E.C. Strategy* (Lexington Books, 1977) and M. Abdel-Fadit (ed.), *Papers on the Economics of Oil* (O.U.P., 1979).

294

2. *The U.S.A. and World Oil*

M. G. de Chazeau and A. E. Kahn's *Integration and Competition in the Petroleum Industry* (Yale U.P., 1959), W. F. Lovejoy and P. T. Homan's *Economic Aspects of Oil Conservation and Regulation* (R.F.F., 1967) and E. W. Zimmerman, *Conservation in the Production of Petroleum* (Yale U.P., 1957) will give even the most avid reader enough background to the very closely studied U.S. oil industry. Supplementary reading is, however, necessary for further background to the significance of U.S. oil import controls for the industry in the late 1950s and the 1960s. See E. H. Shaffer, *The Oil Import Program of the United States* (Praeger, 1968), and G. D. Nash, *United States Oil Policy, 1890–1964* (U. of Pittsburgh Press, 1968) and D. R. Bohi and M. Russell, *Limiting Oil Imports* (Johns Hopkins U.P., 1978). For an interpretation of the more recent U.S. position, when it became more dependent on imports and also seriously affected by environmental considerations, see C. T. Cicchetti, *Alaskan Oil, Alternative Routes and Markets* (Johns Hopkins U.P., 1972), W. J. Mead and A. E. Utton (eds), *U.S. Energy Policy* (Ballinger, 1979) and R. Stobaugh and D. Yergin, *Energy Future* (Random House, 1979). A book by a former senior government official, responsible over many years for investigations into the oil industry, provides a comprehensive analysis of the organization of the United States oil industry and of U.S. involvement in world oil issues. This is J. M. Blair, *The Control of Oil* (Macmillan, 1977). By way of contrast, by virtue of its severe criticism of government intervention in the oil industry, is D. Glasner's *Politics, Prices and Petroleum: the Political Economy of Energy* (Pacific Studies in Public Policy, 1985).

3. *Soviet Oil Development*

Two books by R. M. Campbell provide the most accessible information and analysis on the Soviet oil industry. They are *The Economics of Soviet Oil and Gas* and *Trends in the Oil and Gas Industry* (Johns Hopkins U.P., 1968 and 1976 respectively). D. Park, *Oil and Gas in Comecon Countries* (Kogan Page, London, 1979) is more up to date and includes an analysis of Eastern Europe's position. J. P. Stern's *Soviet Natural Gas Development to 1990* (Lexington Books, 1980) is concerned mainly with natural gas but, of necessity, also has to consider the U.S.S.R.'s oil industry. The report of a NATO colloquium *CMEA; Energy 1980–90* (Brussels, 1982) on the other hand, puts Soviet oil problems and prospects in a wider perspective. For an interpretation of external Soviet policy towards oil, including its oil and gas export policies, see J. Russell, *Energy as a Factor in Soviet Foreign Policy* (Saxon House, 1976), A. J. Klinghoffer, *The Soviet Union and International Oil Politics* (Columbia U.P., 1977), L. Dienes and T. Shabad, *The Soviet Energy System; Resource Use and Policies* (Wiley, 1979) and J. Russell, *Geopolitics of Natural Gas* (Ballinger, 1983). A recent survey of Soviet oil and gas policies in relation to East and West European problems and opportunities is J. P. Stern's *East European Energy and East–West Trade in Energy* (British Institutes Joint Energy Policy Programme, Paper no. 1, London, 1982).

4. *The Major Oil-Exporting Countries*

There is a wealth of reading material on the oil-exporting countries, particularly the Middle East. The general oil industry texts already listed for Chapter 1 pay a great deal of attention to these countries but more specialized work on Venezuela includes E. Lieuwen's *Petroleum in Venezuela: A History* (U. of California Press, 1954), F. Tugwell, *The Politics of Oil in Venezuela* (Stanford U.P., 1975), and L. Vallenilla, *Oil: the Making of a New Economic Order* (McGraw Hill, 1975), a Venezuelan's view of his country's oil industry. G. Coronel, *The Nationalization of the Venezuelan Oil Industry* (Lexington Books, 1983) brings the situation up to date. S. H. Longrigg's *Oil in the Middle East* (O.U.P., 3rd ed., 1968) is a history of discovery and development by an author who was himself involved in the history as a British civil servant, military officer and oil company executive. D. Hirst's *Oil and Public Opinion in the Middle East* (Faber & Faber, 1966) presents the development of Middle Eastern oil from a point of view which contrasts strongly with that in the previous book; C. Issawi and M. Yeganeh's *The Economics of Middle Eastern Oil* (Praeger, 1962) and G. W. Stocking, *Middle East Oil: A Study in Political and Economic Controversy* (Allen Lane, the Penguin Press, 1971), are important books on the evolution of the oil industry in this region. The specific role of Saudi Arabia is dealt with in Sheikh R. Ali, *Saudi Arabia and Oil Diplomacy* (Praeger, 1976). Israel is, of course, much involved with the influences of oil in the Middle East. This is covered in a book by M. Abir, *Oil Power and Politics, Conflict in Arabia* (Cass, 1974) and, more recently, in E. Kanovsky, *The Diminishing Importance of Middle East Oil* (Holmes and Meier, 1982). More generally on the recent geo-politics of Middle East oil, see B. Shwadran, *Middle East Oil: Issues and Problems* (Harvard U.P., 1977) and J. E. Peterson (ed.), *The Politics of Middle Eastern Oil* (Middle East Institute, 1983). A. Hunter's 'The Indonesian Oil Industry', in *Australian Economic Papers*, 5, 1966, O. J. Bee, *The Petroleum Resources of Indonesia* (Oxford U.P., 1982), S. R. Pearson, *Petroleum and the Nigerian Economy* (Stanford U.P., 1970) and F. C. Waddams, *The Libyan Oil Industry* (Croom Helm, 1980) cover other producing areas, though they do not, unfortunately, incorporate the impact of most recent developments. D. Aperys' *The Oil Market in the 1980s* (Harper and Row, 1982) attempts to show how the O.P.E.C. countries can reach a compromise on their oil production policies in the 1980s. Westermann's *Petro-atlas Erdöl und Erdgas* (3rd ed., 1982) locates oil industry activities in all producing countries in maps which are generally presented in English. The accompanying text is, however, in German.

5. *Oil Policies in Western Europe*

Apart from publications on oil by national governments, the European Common Market, the Council of Europe, the Organization for Economic Co-operation and Development and the U.N. Economic Commission for Europe also publish regular studies on oil in Europe. An early E.C.E. document, *The Price of Oil in*

Western Europe (U.N., Geneva, 1955), was one of the most influential in affecting European policies towards the industry. Surprisingly, however, there is not yet a book which deals comprehensively with the rapid growth of the European oil industry in its economic and political environment, but W. G. Jensen, *Energy in Europe, 1945–80* (Foulis, London, 1967), N. J. D. Lucas, *Energy and the European Communities* (Europa Publications, London, 1977) and H. Maull, *Europe and World Energy* (Butterworths, 1980) go part way towards this, and G. W. Hoffman, *The European Energy Challenge, East and West* (Duke U.P., 1985), is up-to-date in its information and interpretations but, unusually, it is concerned with both Western and Eastern Europe. Aspects of the external impact of European energy policies are covered in P. R. Odell, 'Energy Policies in the E.E.C. and their Impact on the Third World', in *Survey of E.E.C. and the Third World* (Hodder & Stoughton, 1981). A European who had a marked impact on the oil industry was Enrico Mattei, who ran E.N.I. until his death: see P. H. Frankel's *Mattei: Oil and Power Politics* (Faber & Faber, 1966). Europe's oil difficulties arising from political and military upheavals in its main supply area were analysed by H. Lubell in *Middle East Oil Crisis and Western Europe's Energy Supplies* (Johns Hopkins U.P., 1963). The analysis of this problem has been brought up to date by E. N. Krapels, *Oil and Security: Problems and Prospects of Importing Countries* (Adelphi Papers no. 136, International Institute for Strategic Studies, London, 1977). For an analysis of the potential impact of indigenous oil and gas production, see P. R. Odell, *The West European Energy Economy: The Case for Self-Sufficiency* (Stenfert Kroese, Leiden, 1976), *The Implications of North Sea Oil and Gas* (eds I. Smart and M. Saeter, Oslo U.P., 1975) and 'Oil and Gas Exploitation in the North Sea', in *Ocean Yearbook 1* (eds E. Borgese and N. Ginsburg, U. of Chicago Press, 1979); K. P. Chapman, *North Sea Oil and Gas* (David and Charles, 1975), D. I. Mackay and G. A. Mackay, *The Political Economy of North Sea Oil* (Robertson, 1975); A. Hamilton, *North Sea Impact* (International Institute for Economic Research, London, 1978). Ø. Noreng, *The Oil Industry and Government in the North Sea* (Croom Helm, 1980) and J. D. Davis, *High-cost Oil and Gas Resources* (Croom Helm, 1981) are specifically concerned with aspects of North Sea developments. W. Molle and E. Wever, *Oil Refineries and Petrochemical Industries in Western Europe: Buoyant Past and Uncertain Future* (Gower Publishing, 1984) are, by contrast, concerned with the fortunes of the continent's downstream oil industry, now in decline.

6. *Japan: Growth and Dependence on Oil Imports*

The author knows of no publications in English which are concerned specifically or mainly with oil in Japan, but the O.E.C.D. does include Japan in its area of study and its publications on oil are sometimes helpful. A short article on the industry in Japan by Y. Tsurumi in the 'Oil Crisis' edition of *Daedalus* (Vol. 104, Fall 1975), Journal of the American Academy of Arts and Sciences, Cambridge, is useful, and Yuan-li Wu's study, *Japan's Search for Oil* (Hoover Institution

Press, Stanford, 1977) is concerned particularly with the international implications of Japan's high degree of dependence on foreign oil.

7. Dependence on Oil in the Developing World

Professor Penrose's book, *The Large International Firm in Developing Countries* (see recommended reading for Chapter 1), must be the first choice for additional reading on the oil industry in the developing world even though it is now over ten years old. M. Tanzer's *The Political Economy of International Oil and the Underdeveloped Countries* (Beacon Press, Boston, 1969) is also concerned with this subject, while P. das Gupta, *The Oil Industry in India* (Allen & Unwin, 1971) presents an important case study. A more recent book on the oil industry in India is R. Vedavalli's *Private Foreign Investment and Economic Development: A Case Study of Petroleum in India* (Cambridge, 1976). *Oil and Politics in Latin America: Nationalist Movements and State Companies* (C.U.P., 1982) by G. Philip analyses oil industry relationships in Latin America. P. R. Odell's 'Energy Prospects for Latin America', in *Bank of London and South America Review*, Vol. 14, No. 2, May 1980, and 'Oil and Gas Potential in Developing Countries and Prospects for its Development', in *Petroleum Exploitation Strategies in Developing Countries* (United Nations, 1982), look at the Third World's opportunities for increasing oil and gas production. Other contributions to the latter – a United Nations sponsored study – cover other aspects of the prospects for oil and gas in the developing countries. The U.N. Economic Commissions for Africa, for Asia and the Far East and for Latin America have been much concerned with oil and have published many surveys. The World Bank's new Energy Division has also surveyed Third World oil and gas potential in recent reports as, for example, in *Energy in the Developing Countries* (World Bank, Washington, 1980) and in its more recent study, *The Energy Transition in Developing Countries* (1983).

8. Oil in International Relations and World Economic Development

Most of the books recommended for earlier chapters also provide background material to ideas introduced in this chapter. But see also G. Lenczowski's *Oil and State in the Middle East* (Cornell U.P., 1960) and his brief but much more recent *Middle East Oil in a Revolutionary Age* (American Enterprise Institute for Public Policy Research, 1976). B. Shwadran's *The Middle East, Oil and the Great Powers* (New York, 1959) has now been supplemented by the author's new book on Middle East oil and listed under the readings for Chapter 4. Publications of the Organization of Petroleum Exporting Countries (Vienna) give the O.P.E.C. view and this is critically examined in F. Ghadar's *The Evolution of O.P.E.C. Strategy* (Levington, 1977). S. H. Schurr and P. T. Homan, *Middle East Oil and the Western World* (American Elsevier Publishing Co., New York, 1971), provide a detailed and systematic analysis of the economic implications of the dominant role of Middle East oil in the world oil economy pre-1973. N. S. Houthakker's

Suggestions for Further Reading

The World Price of Oil: A Medium Term Analysis (American Enterprise Institute for Public Policy Research, 1976), Y. S. Park's *Oil Money and the World Economy* (Wilton House Publications, London, 1976) and T. M. Rybezynski's (ed.) *The Economies of the Oil Crisis* (Macmillan, 1976) attempt to bring the story up to date. M. Bailey's *Oilgate: the Sanctions Scandal* (Hodder & Stoughton, 1979) deals with the relationships of the international oil companies with Rhodesia and South Africa in the context of international sanctions against the former country. World Bank reports on Venezuela, Libya, Kuwait, etc., clearly demonstrate the importance of oil revenues in the economic development of major producing countries, for many of which analyses of the oil/economic development relationship have been made. See, for example, J. Salazar-Casillo, *Oil in the Economic Development of Venezuela* (New York, 1976). J. A. Allan, *Libya: the Experience of Oil* (Croom Helm, 1982) shows how the development of the oil industry affected the traditional activities and society of Libya. The important question of U.S. relations with its oil neighbour is examined in J. R. Ladman *et al* (eds), *U.S.–Mexican Energy Relationship* (Lexington Books, 1981).

9. The Revolution in the World of Oil Power, 1970–75

The changes in the world oil situation described and analysed in this chapter are still very recent but there is, nevertheless, already a considerable literature on the subject, so reflecting the importance of the changes that have taken place. For early contrasting views on the nature of the oil crisis in 1973–4 see the following articles in successive issues of the American journal *Foreign Affairs*: M. A. Adelman, 'Is the Oil Crisis Real?', Winter 1972–3; J. E. Akins, 'This Time the Wolf Really is at the Door', Spring 1973; J. Amuzegar, 'The Oil Story: Facts, Fiction and Fair Play', Summer 1973. And for an interpretation of the mid-1975 position see P. R. Odell, 'The World of Oil Power in 1975', in *The World Today*, Royal Institute of International Affairs, London, July 1975. N. Choucri, *International Politics of Energy Interdependence* (Lexington Books, 1976), F. R. Wyant, *The United States, O.P.E.C. and Multinational Oil* (Lexington Books, 1977), R. Mabro (ed.), *World Energy: Issues and Policies* (O.U.P., 1980) and R. El Mallakh, *O.P.E.C.: Twenty Years and Beyond* (Westview, 1981) offer more recent interpretations.

10. The Response to O.P.E.C.'s Challenge

The prospects for the future availability of oil constitute a main input into analyses on the ability to challenge O.P.E.C.'s recent domination of the world market. These are being seriously debated. For contrasting views see, *Energy: Global Prospects 1985–2000*, Reports of the Workshop on Alternative Energy Strategies (McGraw-Hill, New York, 1977); *The Future Supply of Nature-made Petroleum and Gas*, ed. R. F. Meyer (Pergamon Press, 1977); P. L. Eckbo, *The Future of World Oil* (M.I.T. Press, Cambridge, 1975); P. R. Odell, *Energy: Needs and*

Resources (Macmillan, 2nd ed., 1977) and P. R. Odell and K. E. Rosing, *The Future of Oil* (Kogan Page, 2nd ed., 1983). On the prospects for oil from China, see C.-Y. Cheng, *China's Petroleum Industry* (New York, 1976) and H. C. Ling, *The Petroleum Industry of the People's Republic of China* (Stanford U.P., 1975). References to U.S., U.S.S.R., Mexican, North Sea and Third World oil supply prospects have been listed under the chapters concerned with these countries. The prospects for natural gas, both complementary and competitive with oil, are evaluated in J. D. Davis, *Blue Gold: the Political Economy of Natural Gas* (Allen and Unwin, 1984). For an analysis which assumes that oil is a scarce commodity and investigates the implications of this for the structure of energy intensive, industrialized countries see, G. T. Goodman *et al* (eds), *The European Transition from Oil* (Academic Press, 1981).

Similarly, the prospects for the future of oil demand are being critically re-examined in the light of recent experience which indicates that oil products can be used much more efficiently than hitherto and more easily substituted by other sources of energy than previously supposed. The I.E.A.'s recent *World Energy Outlook* (O.E.C.D., Paris, 1982), updated in the organization's *Energy Review 1984*, includes a detailed analysis on this development. B. Fritsch's *Energy Demand of Industrialized and Developing Countries to 1990* (Centre for Economic Research, Institute of Technology, Zürich, 1982) is concerned with the same issue.

For the shorter term the oil supply/demand situation remains characterized by uncertainty. This is examined in P. W. MacAvoy, *Crude Oil Prices* (Ballinger, 1982) and P. K. Verleger, Jr, *Oil Markets in Turmoil* (Ballinger, 1982) and, in respect of security of supply issues specifically, by E. N. Krapels, *Oil Crisis Management, Strategic Stockpiling for International Security* (J. Hopkins U.P., 1980).

On the wider issues of the future role of oil in the international political and economic system and of the relationships between the parties involved, see P. R. Odell and L. Vallenilla, *The Pressures of Oil: A Strategy for Economic Revival* (Harper & Row, 1978) and T. Hoffman and B. Johnson, *The World Energy Triangle: a Strategy for Cooperation* (Ballinger, 1981). The Brandt Commission Report, *North-South: A Programme for Survival* (Pan Books, 1980) is concerned in part with the impact of oil on international relationships between the industrialized countries and the Third World. The world as seen from the perspective of the oil exporters is presented in J. Evans, *O.P.E.C., its Member States and the World Energy Market* (Longman, 1985).

Postscript: on Keeping Up to Date

Readers will have become aware just how dynamic the oil industry has become over the last decade and a half and may well want to keep up to date with events and developments which influence the issues and relationships with which this book has been concerned. Indispensable from this point of view is the monthly publication *Petroleum Economics* (available on subscription only from P.O. Box

105, 25–31 Ironmonger Row, London EC1U 3PN, but also available in many public libraries).

With the background knowledge and understanding of the industry, acquired as a result of reading this book, *Petroleum Economics* will make sense and it will provide each month an immense amount of information. One word of warning, however: it more often than not reflects conventional wisdom in general, and the views of the major international oil companies in particular and it does not, therefore, usually present a comprehensive range of opinion and interpretations on many important issues which currently affect the oil industry, particularly those on the relationships of the international oil companies with governments and inter-governmental organizations. Unfortunately there is no complementary, radical journal devoted to these issues, though such articles do appear in other journals from time to time and, as indicated in the preceding pages, there are many books which are devoted to a more radical analysis of world oil. Three specialist quarterly journals devoted, in part, to oil questions are *Energy Policy* (Butterworth Scientific Press, Journals Division, Guildford), *Natural Resources Forum* (a publication of the U.N. Centre for Natural Resources, New York) and *Energy Exploration and Exploitation* (Elsevier Applied Science Publications, Barking). The *Energy Journal*, the main publication of the International Association of Energy Economists (Washington, D.C. and affiliates in many countries) is also much concerned with oil issues.

Index

Index

Index

Index

Index

Index

Index

MORE ABOUT PENGUINS, PELICANS, PEREGRINES AND PUFFINS

For further information about books available from Penguins please write to Dept EP, Penguin Books Ltd, Harmondsworth, Middlesex UB7 0DA.

In the U.S.A.: For a complete list of books available from Penguins in the United States write to Dept DG, Penguin Books, 299 Murray Hill Parkway, East Rutherford, New Jersey 07073.

In Canada: For a complete list of books available from Penguins in Canada write to Penguin Books Canada Limited, 2801 John Street, Markham, Ontario L3R 1B4.

In Australia: For a complete list of books available from Penguins in Australia write to the Marketing Department, Penguin Books Australia Ltd, P.O. Box 257, Ringwood, Victoria 3134.

In New Zealand: For a complete list of books available from Penguins in New Zealand write to the Marketing Department, Penguin Books (N.Z.) Ltd, Private Bag, Takapuna, Auckland 9.

In India: For a complete list of books available from Penguins in India write to Penguin Overseas Ltd, 706 Eros Apartments, 56 Nehru Place, New Delhi 110019.

A CHOICE OF
PELICANS AND PEREGRINES

☐ *The Knight, the Lady and the Priest*
Georges Duby

The acclaimed study of the making of modern marriage in medieval France. 'He has traced this story – sometimes amusing, often horrifying, always startling – in a series of brilliant vignettes' – *Observer*

☐ *The Limits of Soviet Power* **Jonathan Steele**

The Kremlin's foreign policy – Brezhnev to Chernenko, is discussed in this informed, informative 'wholly invaluable and extraordinarily timely study' – *Guardian*

☐ *Understanding Organizations* **Charles B. Handy**

Third Edition. Designed as a practical source-book for managers, this Pelican looks at the concepts, key issues and current fashions in tackling organizational problems.

☐ *The Pelican Freud Library: Volume 12*

Containing the major essays: *Civilization, Society and Religion, Group Psychology* and *Civilization and Its Discontents*, plus other works.

☐ *Windows on the Mind* **Erich Harth**

Is there a physical explanation for the various phenomena that we call 'mind'? Professor Harth takes in age-old philosophers as well as the latest neuroscientific theories in his masterly study of memory, perception, free will, selfhood, sensation and other richly controversial fields.

☐ *The Pelican History of the World*
J. M. Roberts

'A stupendous achievement . . . This is the unrivalled World History for our day' – A. J. P. Taylor

A CHOICE OF
PELICANS AND PEREGRINES

☐ **A Question of Economics** **Peter Donaldson**

Twenty key issues – from the City and big business to trades unions – clarified and discussed by Peter Donaldson, author of *10 × Economics* and one of our greatest popularizers of economics.

☐ **Inside the Inner City** **Paul Harrison**

A report on urban poverty and conflict by the author of *Inside the Third World*. 'A major piece of evidence' – *Sunday Times*. 'A classic: it tells us what it is really like to be poor, and why' – *Time Out*

☐ **What Philosophy Is** **Anthony O'Hear**

What are human beings? How should people act? How do our thoughts and words relate to reality? Contemporary attitudes to these age-old questions are discussed in this new study, an eloquent and brilliant introduction to philosophy today.

☐ **The Arabs** **Peter Mansfield**

New Edition. 'Should be studied by anyone who wants to know about the Arab world and how the Arabs have become what they are today' – *Sunday Times*

☐ **Religion and the Rise of Capitalism**
   R. H. Tawney

The classic study of religious thought of social and economic issues from the later middle ages to the early eighteenth century.

☐ **The Mathematical Experience**
   Philip J. Davis and Reuben Hersh

Not since *Gödel, Escher, Bach* has such an entertaining book been written on the relationship of mathematics to the arts and sciences. 'It deserves to be read by everyone ... an instant classic' – *New Scientist*

A CHOICE OF PENGUINS

☐ **The Complete Penguin Stereo Record and Cassette Guide**
Greenfield, Layton and March

A new edition, now including information on compact discs. 'One of the few indispensables on the record collector's bookshelf' – *Gramophone*

☐ **Selected Letters of Malcolm Lowry**
Edited by Harvey Breit and Margerie Bonner Lowry

'Lowry emerges from these letters not only as an extremely interesting man, but also a lovable one' – Philip Toynbee

☐ **The First Day on the Somme**
Martin Middlebrook

1 July 1916 was the blackest day of slaughter in the history of the British Army. 'The soldiers receive the best service a historian can provide: their story told in their own words' – *Guardian*

☐ **A Better Class of Person** **John Osborne**

The playwright's autobiography, 1929–56. 'Splendidly enjoyable' – John Mortimer. 'One of the best, richest and most bitterly truthful autobiographies that I have ever read' – Melvyn Bragg

☐ **The Winning Streak** **Goldsmith and Clutterbuck**

Marks & Spencer, Saatchi & Saatchi, United Biscuits, GEC . . . The UK's top companies reveal their formulas for success, in an important and stimulating book that no British manager can afford to ignore.

☐ **The First World War** **A. J. P. Taylor**

'He manages in some 200 illustrated pages to say almost everything that is important . . . A special text . . . a remarkable collection of photographs' – *Observer*

A CHOICE OF PENGUINS

☐ *Man and the Natural World* **Keith Thomas**

Changing attitudes in England, 1500–1800. 'An encyclopedic study of man's relationship to animals and plants . . . a book to read again and again' – Paul Theroux, *Sunday Times* Books of the Year

☐ *Jean Rhys: Letters 1931–66*
 ·Edited by Francis Wyndham and Diana Melly

'Eloquent and invaluable . . . her life emerges, and with it a portrait of an unexpectedly indomitable figure' – Marina Warner in the *Sunday Times*

☐ *The French Revolution* **Christopher Hibbert**

'One of the best accounts of the Revolution that I know . . . Mr Hibbert is outstanding' – J. H. Plumb in the *Sunday Telegraph*

☐ *Isak Dinesen* **Judith Thurman**

The acclaimed life of Karen Blixen, 'beautiful bride, disappointed wife, radiant lover, bereft and widowed woman, writer, sibyl, Scheherazade, child of Lucifer, Baroness; always a unique human being . . . an assiduously researched and finely narrated biography' – *Books & Bookmen*

☐ *The Amateur Naturalist*
 Gerald Durrell with Lee Durrell

'Delight . . . on every page . . . packed with authoritative writing, learning without pomposity . . . it represents a real bargain' – *The Times Educational Supplement*. 'What treats are in store for the average British household' – *Daily Express*

☐ *When the Wind Blows* **Raymond Briggs**

'A visual parable against nuclear war: all the more chilling for being in the form of a strip cartoon' – *Sunday Times*. 'The most eloquent anti-Bomb statement you are likely to read' – *Daily Mail*

PENGUIN REFERENCE BOOKS

☐ *The Penguin Map of the World*

Clear, colourful, crammed with information and fully up-to-date, this is a useful map to stick on your wall at home, at school or in the office.

☐ *The Penguin Map of Europe*

Covers all land eastwards to the Urals, southwards to North Africa and up to Syria, Iraq and Iran * Scale = 1:5,500,000 * 4-colour artwork * Features main roads, railways, oil and gas pipelines, plus extra information including national flags, currencies and populations.

☐ *The Penguin Map of the British Isles*

Including the Orkneys, the Shetlands, the Channel Islands and much of Normandy, this excellent map is ideal for planning routes and touring holidays, or as a study aid.

☐ *The Penguin Dictionary of Quotations*

A treasure-trove of over 12,000 new gems and old favourites, from Aesop and Matthew Arnold to Xenophon and Zola.

☐ *The Penguin Dictionary of Art and Artists*

Fifth Edition. 'A vast amount of information intelligently presented, carefully detailed, abreast of current thought and scholarship and easy to read' – *The Times Literary Supplement*

☐ *The Penguin Pocket Thesaurus*

A pocket-sized version of Roget's classic, and an essential companion for all commuters, crossword addicts, students, journalists and the stuck-for-words.